ALASKA'S FARMS and GARDENS

Volume 11, Number 2, 1984
ALASKA GEOGRAPHIC®

To teach many more to better know and use our natural resources

Editor: Penny Rennick
Associate Editor: Kathy Doogan
Designer: Sandi Harner

About This Issue:

Alaska's Farms and Gardens is the result of several years' efforts by our staff to gather information on the state's agricultural efforts, big and small. Terrence Cole, editor of *The ALASKA JOURNAL®*, contributed a historical perspective with his review of early-day agriculture in Alaska. We are grateful for the important assistance of the Cooperative Extension Service; also of the late Curtis Dearborn, and of Don Dinkel, now retired, researchers with the Alaska Agricultural Experiment Station; of Carl Amstrup, Executive Director, Alaska Agricultural Action Council; and of Delon Brown, Alaska Crop and Livestock Reporting Service. We also thank Ken Kreig, Walt Them, and Wayne Vandre of the Cooperative Extension Service for reviewing portions of the manuscript.

Library of congress cataloging in publication data
Main entry under title:
Alaska's farms and gardens.
(Alaska geographic ; v. 11, no. 2)
1. Agriculture—Alaska. 2. Gardening—Alaska.
I. Alaska Geographic Society. II. Series.
F901.M266 vol. 11, no. 2 917.98s 630'.9798 84-6455
[S507.A4]
ISBN 0-88240-202-1

Cover — *The vegetable garden of Dave and Eleanor Young is a nice complement to their Cooper Landing log home. (Helen Rhode)*

Previous page — *A midsummer display from Alf and Edna Kalvick's well-tended garden in Skagway demonstrates how productive an Alaska garden can be. Clockwise, from lower left, are: radishes, celery, carrots, head lettuce, onions, beets, parsley, and peas. (Bob Coffey)*

ALASKA GEOGRAPHIC®,ISSN 0361-1353, is published quarterly by The Alaska Geographic Society, Anchorage, Alaska 99509-6057. Second-class postage paid in Edmonds, Washington 98020-3588. Printed in U.S.A. Registered trademark: Alaska Geographic. ISSN 0361-1353; Key title Alaska Geographic.

THE ALASKA GEOGRAPHIC SOCIETY is a nonprofit organization exploring new frontiers of knowledge across the lands of the polar rim, learning how other men and other countries live in their Norths, putting the geography book back in the classroom, exploring new methods of teaching and learning — sharing in the excitement of discovery in man's wonderful new world north of 51° 16'.

MEMBERS OF THE SOCIETY RECEIVE *ALASKA GEOGRAPHIC®*, a quality magazine which devotes each quarterly issue to monographic in-depth coverage of a northern geographic region or resource-oriented subject.

MEMBERSHIP DUES in The Alaska Geographic Society are $30 per year; $34 to non-U.S. addresses. (Eighty percent of each year's dues is for a one-year subscription to *ALASKA GEOGRAPHIC®*.) Order from The Alaska Geographic Society, Box 4-EEE, Anchorage, Alaska 99509-6057; (907) 274-0521.

MATERIAL SOUGHT: The editors of *ALASKA GEOGRAPHIC®* seek a wide variety of informative material on the lands north of 51° 16' on geographic subjects — anything to do with resources and their uses (with heavy emphasis on quality color photography) — from Alaska, northern Canada, Siberia, Japan — all geographic areas that have a relationship to Alaska in a physical or economic sense. We do not want material done in excessive scientific terminology. A query to the editors is suggested. Payments are made for all material upon publicaton.

CHANGE OF ADDRESS: The post office does not automatically forward *ALASKA GEOGRAPHIC®* when you move. To ensure continuous service, notify us six weeks before moving. Send us your new address and zip code (and moving date), your old address and zip code, and if possible send a mailing label from a copy of *ALASKA GEOGRAPHIC®*. Send this information to *ALASKA GEOGRAPHIC®* Mailing Offices, 130 Second Avenue South, Edmonds, Washington 98020-3588.

MAILING LISTS: We have begun making our members' names and addresses available to carefully screened publications and companies whose products and activities might be of interest of you. If you would prefer not to receive such mailings, please so advise us, and include your mailing label (or your name and address if label is not available).

Alaska's Farm and Pasture Lands

Yellow indicates potential farm and pasture lands; green indicates areas where commercial agriculture is taking place.

RAN

Contents

Spectators gather around a display of record-sized vegetables at the 1982 Alaska State Fair at Palmer. (Bruce Holser)

Jamboree Stormy (left) and Bacardi Fizz, both registered Arabian stallions, get acquainted at their home, the Ranch of Envy in Palmer. (Jane Gnass)

Hay, baled for winter storage, dots a field in the Matanuska Valley. Snow-dusted Pioneer Peak rises in the background. (Third Eye Photography)

This volume is not intended to be a bible on how to raise things successfully in Alaska, but merely an overview to portray, in a good collection of photos with brief text, what the farmer, gardener, and livestock rancher might expect in Alaska. At the back of this volume there is a list of addresses to which one may write for more specific information on land (old homestead laws are no more and state, federal, and now Native corporations have limited procedures for acquiring land).

As to the great glorious future of agriculture in Alaska . . . it is still exciting to know we in Alaska are able to grow many things . . . even brag about their unique qualities as well as sometimes great size. Long daylight hours do pump in a lot of goodies during the short growing season . . . if a sneaky frost doesn't come along. But land to own, and land that will grow crops well, is as limited as the markets . . . each community in some degree insulated by distance from its neighboring communities . . . and wages and the costs of labor are high. But so are the freight rates from "The Outside" which will always guarantee some market for some locally produced farm goods.

In southeastern Alaska, the long panhandle extending down along the Coast Mountains to within five hundred miles of the continental United States border . . . it's mild . . . and wet . . . a condition that exists in less mild and often wetter degree around the rim of Prince William Sound.

The more "farmable" Alaska begins on the Kenai Peninsula where some Interior weather influence is felt and there is some shelter from the Gulf of Alaska's heavier precipitation. The most "farm country" is north of there in the Matanuska Valley and north of there in the Tanana Valley where Interior temperatures are much higher in summer, much colder in winter.

There are "good farming" pockets here and there, but that gives you a rough idea . . . some hopes of grain exports, modest hopes for local markets . . . but Alaska is not likely to be "big" farm country. Even with high freight costs, it is generally cheaper to ship things in from Outside than it is to raise and sell them locally.

But we still remember as a boy at the old Southeast Alaska Fairs in Juneau, seeing the Haines strawberries displayed, one to a demitasse cup, sitting red and proud above the rim . . . the big fat carrots that weighed several pounds. The huge turnips and the monster cabbage . . . all sorts of fine potatoes . . . scads of berries and a riotous color display of an infinite variety of flowers. Did you know a dahlia from the Blanchard Gardens in Skagway used to annually win the Pacific Coast Dahlia Exposition at San Francisco . . . or that pansies in that garden went five inches across the face?

What we mean is, Alaska and the fruits of her gardens and fields are to brag about. Some folks even make a living at them.

Robert A. Henning

President
The Alaska Geographic Society

Right — *Record-sized cabbages were a family affair at the 1983 Alaska State Fair at Palmer. The blue ribbon went to 15-year-old Mark Dinkel of Wasilla, whose 83½-pound entry set a new world record. Mark's father, Gene, took second place with a hefty 82-pounder.* *(Bruce Holser)*

Far right — *Straight, neat rows of potatoes, the state's most important vegetable crop, ripen in the Matanuska Valley in August. Nearly $2 million worth of potatoes were raised in Alaska in 1982.* *(Louise Franzmann)*

Introduction

The popular image of Alaska as a land of perpetual ice and snow has long made it seem to some people like a preposterous place for farming. When the United States purchased Alaska in 1867, *Harper's Weekly* imagined the unusual farms that a traveler might find in "Seward's Icebox." The magazine explained that the fields in Alaska were thick every season with fresh crops of ice, and that Alaskan cattle all naturally produced ice cream.

There is really little need to dwell on the fact that when Alaskan farmers milk their cows, they do not fill up their pails with vanilla ice cream. But most writers about agriculture in Alaska fell into that trap. In the 1940s, George Sundborg carefully examined the literature on Alaskan agriculture, and found that much

Left — *Horses forage on Near Island, just offshore from downtown Kodiak. (Perry Valley)*

Below — *Beef cattle, mostly polled Herefords, are herded into corrals at Bruce Willard's cattle camp at the head of Kachemak Bay. Willard Farms, of Homer, runs the largest beef cattle herd and operates the only certified slaughterhouse on the lower Kenai Peninsula. (Janet and Robert Klein)*

Fascinated by a feeding sow and her piglets, a visiting toddler cautiously approaches the animals on Charlotte and Fred Boden's Anchor Point hog ranch. Domestic animals, as with their wild counterparts, are unpredictable; parents should teach their children a healthy respect for all creatures. (Janet and Robert Klein)

of it was from the "believe it or not, plants actually grow in Alaska" school. As Sundborg wrote, "to refute the 'Alaska-is-a-land-of-ice-and-snow' legend, the point has been belabored unmercifully. If the simple fact could have been taken for granted that the laws of nature operate in Alaska just as they do everywhere else, writers on the subject might have gone on to tell more of the when, where, how, why, and what of farming in Alaska."

This book attempts to do just that. While Alaskan farmers and gardeners do face many difficult climatic and environmental problems that are not of concern in lands farther to the south, and they must often learn new techniques of tilling the soil, it is no miracle to grow potatoes or broccoli in Alaska.

Since the Russian fur traders started expanding into Alaska in the eighteenth century, most of the food for the territory has been imported from Outside. Today the major part of all the foods sold in Alaska's grocery stores are shipped north thousands of miles from the Lower 48 by airplane, barge, or truck, enabling Alaskan families to eat California vegetables all year-round. Cold storage produce, however, is expensive, and it is often bland and flavorless, so a small garden in Alaska can be especially valuable.

But there is more to gardening than producing a food crop and saving money on groceries. Gardening can be a relaxing, soul-satisfying hobby, with its own unique rewards. As Mike Seniszuk, an experienced gardener from Dawson City, put it, "Hell, who cares about prices? I like to see things grow."

Most of the pages of this book tell how to plan for, plant, maintain, and harvest a garden in Alaska. Other sections examine large-scale commercial agriculture in Alaska, especially livestock and the production of grains. Though farming and gardening have been practiced in the north for hundreds of years, Alaskan agriculture is still in the experimental stage, as new adaptations and techniques are constantly being developed. As Bulletin #1 of the Alaska Agricultural Experiment Stations warned in 1902, "The pioneer who comes to Alaska to farm often finds that it is not safe to follow the customs with which he is familiar in regions farther south."

For your first Alaska garden, heed the basic how-tos of the experts. But as you gain confidence and experience at gardening on the last frontier you might want to try some experiments of your own. Expect some failure, but remember that some of the biggest, most tender vegetables in the world can be grown in Alaska.

Rhode Island reds (brown), barred Plymouth Rock hens (black and white), and a white leghorn gather in their coop in the Matanuska Valley. In 1982, the valley's 35,700 hens and pullets of laying age produced six million eggs. (Jane Gnass)

Although Alaska's growing season is short, long daylight hours and mild temperatures result in some exceptional vegetables. Greg Norris (above) shows off several giant turnips; right, an unidentified young lady peers cautiously into the depths of a mammoth cabbage. (Both photos by Helen Rhode)

Just picked, these plump red radishes are hosed down with cool water to keep them fresh. (Bob Cellers)

At left, a combine makes its way down a row of peas on a Matanuska Valley farm. The machine removes the pods, leaving vines and pods in the field, and stores the peas in a hopper on top of the combine. A full hopper, shown below, holds fifteen hundred pounds of peas.
(Both photos by Bob Cellers)

A conveyor carries potatoes from field to truck at Wild Wing Farm near Palmer. Potatoes are the main crop in the Matanuska Valley, and are sold in supermarkets throughout the state.
(Bob Cellers)

Mike Wessel rakes hay to be baled in a field near Palmer, in the Matanuska Valley. In 1982, about one-third of the total hay production in the state occurred in the Matanuska Valley. (Jane Gnass)

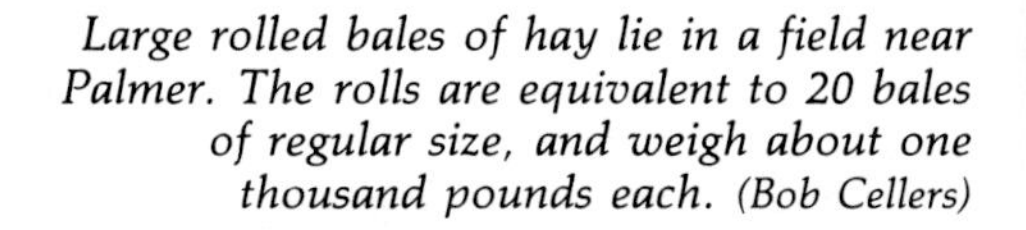

Large rolled bales of hay lie in a field near Palmer. The rolls are equivalent to 20 bales of regular size, and weigh about one thousand pounds each. (Bob Cellers)

A farm house and barn near Palmer make a picturesque scene against the majestic Chugach Mountains. (Jane Gnass)

Right — *Peas proved an easy crop to grow in Fairbanks in 1921. Residents of the Interior depended on their gardens for a year-round vegetable supply. (Charles Bunnell Collection, University of Alaska, Fairbanks)*

Far right — *Gardens provided an essential source of food for early-day settlers of Alaska. In this 1917 photograph, the Louis Spitz family of Ruby, a village on the Yukon River about 90 miles south of the Arctic Circle, stood proudly by their carefully tended vegetable garden. (Harry M. DeVane Collection, University of Alaska, Fairbanks)*

History of Agriculture in Alaska

Agriculture has a long history in the North dating back to at least the eighteenth century, when the Russians started to grow their own food in Alaska so the fur traders would not have to eat fish all the time. But they were discouraged by the soggy, acid soils along Alaska's coastline. The Russians seldom had bread or flour because they had great difficulty getting grains to ripen. Their vegetable gardens, however, were relatively successful. Leafy plants like cabbage and lettuce tended to be watery, while roots and tubers did very well. The *promyshlenniki* enjoyed fresh turnips, potatoes, radishes, beets, carrots, rutabagas, onions, and garlic throughout Russian America.

In addition to small kitchen gardens, almost every community had a few farm animals, including

chickens, cows, sheep, pigs, horses, and goats. The meat of the pigs and chickens tasted and smelled like fish, their main diet. Descendants of the Russian cattle herds survived in Alaska until the twentieth century, but even on Kodiak Island, the first center of agriculture in Alaska, Russian efforts to raise vegetables and livestock were only marginally successful.

After the United States purchased Alaska in 1867, the major part of Alaska's food supply was imported from Outside. Most of the thousands of gold stampeders who rushed to Alaska at the turn of the century brought at least a year's worth of food with them. Living on a steady diet made from beans, flour, and dehydrated vegetables like potatoes and onions, hungry miners, roadhouse keepers, and restaurant men saw the advantages of trying to grow their own vegetables as the Russians had done. If an early-day Alaskan did not plant a small vegetable garden near his claim or a few radishes and some lettuce in the sod roof of his log cabin, he might have to go without tasting fresh produce for years at a time.

The population of Alaska increased dramatically during the gold rushes, and so did interest in promoting Alaskan agriculture. Transportation routes to the northern gold camps were long and difficult, and it was far cheaper to grow food locally than to carry it hundreds of miles across the wilderness by sternwheeler or dog sled. In 1897 Walter H. Evans of the U.S. Department of Agriculture headed a survey team that visited practically every community throughout Alaska, and reported on the great agricultural potential of the territory. In the wake of this investigation, the federal government established Agricultural Experiment Stations at locations throughout Alaska, including Kodiak, Sitka, Kenai, Rampart, Copper Center, Fairbanks, and in the Matanuska Valley.

The Agricultural Experiment Stations blazed many trails for Alaskan farmers by introducing new strains of crops and hybrid livestock, and developing the specialized techniques needed to make agriculture in Alaska practical. In their early years, the stations experimented with such things as Japanese roses, strawberries, Siberian spring wheat, Galloway-Holstein cattle, honeybees, and Jerusalem artichokes. As a central clearing house for agricultural information gathered by farmers throughout Alaska, the experiment stations performed an invaluable service.

As part of their ongoing research, each year the stations distributed free vegetable seeds and plants to anyone who asked for them. In return they asked that the gardeners write back and explain what worked and what did not work in their fields that season. Many of these letters from Alaska gardeners were reprinted in the annual reports of the experiment stations. "They are in a way more valuable than reports from the experiment stations," the 1907 bulletin admitted, "in that they record the things the people accomplish."

J.K. Calbreath, a miner from Wiseman on the Koyukuk River, far above the Arctic Circle, explained how valuable a small garden could be to a miner in the Alaskan wilderness. "I wish to say that all the seeds sent were good, especially the Petrowski turnip seed," he wrote in 1917. "I sowed all I had and raked them over and had a great crop, the roots weighing from one-half to two pounds. I never witnessed anything

Before the advent of power tools, Owen Meals of Valdez used dog power to hill his potato patch. Valdez, with its rocky, wet soils, is considered one of the more difficult areas of Alaska to garden. (P.S. Hunt photo, Mary Whalen Collection, University of Alaska, Fairbanks)

OWEN MEALS "HILLING POTATOES" VALDEZ, ALASKA.

Above — *Early-day Alaska missions had to be as self-sufficient as possible, as no one could predict with certainty when the next river boat of supplies would arrive. By 1903, the mission at Holy Cross had an 8-acre vegetable garden, 30 acres of grain, and 7 horses. (University Relations Collection, University of Alaska, Fairbanks)*

Left — *Beautiful cucumbers and tomatoes crowded this greenhouse in the early 1900s in the bustling gold mining town of Iditarod. In those days — and sometimes even today — the only source of fresh produce in remote Alaska communities was what the residents could raise themselves. (Archie Lewis Collection, University of Alaska, Fairbanks)*

Left — *In contrast to their surroundings of tree stumps, outhouses, and cabins of rough-hewn lumber, these Valdez gardeners provide a bit of order with their carefully manicured lettuce patch. (P.S. Hunt photo, Mary Whalen Collection, University of Alaska, Fairbanks)*

Below — *Dr. C.C. Georgeson, the man primarily responsible for establishing agricultural experiment stations in many locations in Alaska at the turn of the century, poses with an apple tree at the Sitka station. (University Relations Collection, University of Alaska, Fairbanks)*

Above — *Agriculture in Alaska has often required determination and ingenuity. Here a cow hesitantly surveys her bush plane transportation prior to a 1939 flight from Anchorage to McGrath. (University Relations Collection, University of Alaska, Fairbanks)*

Right — *The Arctic Greenhouse display at the Tanana Valley Fair, probably photographed prior to 1920, was truly a feast for the eyes of Alaskans hungry for a look at fresh produce. The exhibit included celery, leeks, radishes, cucumbers, gourds, tomatoes, eggplant, lettuce, and cauliflower. (Bunnell Collection, University of Alaska, Fairbanks)*

like it, and the best of all, our dogs are very fond of them. Man feed, dog feed, or any other kind of grub here is awfully high — bacon, 75¢; flour, $24.00 per hundred; cornmeal, $2.50 for a 10-pound sack, and other things in proportion. So you see we have all got to dig gardens now in order to prospect." James Minano, a Japanese miner and gardener from Coldfoot, had been in the Arctic since the 1890s. He planted a large garden in the Brooks Range every summer. "Last year we raised close to four tons of potatoes," he wrote in 1918, "besides turnips, cabbage, celery, and other garden truck. I can truthfully say that potatoes averaged a heaping gold pan to the hill."

There were small farms and gardens in nearly every community in Alaska, and the part-time farmers were especially proud of the crops they raised in the land of the midnight sun. "I beg to venture the assertion that no better radishes and turnips can be grown any place in the world than right here in Nome," Judge C.W. Thornton claimed in 1907. But the spirit of Alaska gardeners was probably best summed up by a man who wrote from Kotzebue in 1917. "Everybody around here is garden mad," he said, "and each fellow says he will have the best garden next year."

COMMERCIAL FARMING IN ALASKA

Until the 1930s the agricultural heart of Alaska was in the Tanana Valley, near Fairbanks, where farmers produced potatoes, cabbage, hay, grain, milled flour, and milk, as well as raising hogs and poultry. The most important commercial vegetable crop was the potato, and Tanana Valley farmers produced hundreds of tons for the local market every year.

ARCTIC GREENHOUSE
CHARLES BLASER-PROP.
CRIMSON
GOURDS
CUCUMBERS
LEEK ROUEN

VIEW OF MATANUSKA VALLEY FROM HIKEY FARM: MILE 6 GOV'T R. ROAD.

One of the most successful farmers was Paul Rickert, who had about 100 acres under cultivation just outside Fairbanks, with about 20 acres in vegetables and several large greenhouses. Rickert sold milk, butter, eggs, and pork in the markets of Fairbanks, but he specialized in produce. In his greenhouses he raised tomatoes and cucumbers, for which Fairbanks residents often paid top dollar. In 1908 Rickert grew 2½ tons of greenhouse tomatoes, and raised 15,000 heads of cabbage, 15,000 stalks of celery, 10 tons of potatoes, and large quantities of turnips, beets, carrots, parsnips, peas, and beans.

The success of farmers like Paul Rickert and others made the future of large-scale farming in Alaska seem bright. C.C. Georgeson, the "father of Alaskan agriculture," who headed the Agricultural Experiment Stations for many years, predicted in 1909 that Alaska's farms would one day support a population of from three million to six million people. Rickert believed, however, that for Alaska's population to grow, more Alaskan farmers would have to get married. A common complaint was that most Alaskan farmers were usually "old, broken down prospectors," who were single. These bachelor farmers, it was claimed, only staked homesteads and put up shacks because it was cheaper than living in a hotel during the winter.

Rickert, who helped organize the Tanana Valley Fair Association's first annual fair in 1924, thought that one reason why there were not more farmers in Alaska was that there were more "farmers than farmers' wives and families." He planned to put an end to the imbalance at the 1924 fair. Along with the usual exhibits of livestock, vegetables, cooking, and arts and crafts, Rickert planned a mass wedding ceremony at which he hoped all the single farmers in Fairbanks would get married.

"Get married at the Tanana Valley Fair," he announced a week before the fair opened. "We will furnish the License, the Preacher, the Ring, the Wedding, the Music, and the Flowers . . . Address all applications to P.J. Rickert, Tanana Valley Fair Association Inc."

Despite such a generous offer, and the extra bonus of a "free honeymoon trip by airship" for the lucky couples, no one accepted matchmaker Rickert's offer.

With construction of the Alaska Railroad between 1915 and 1923, the Matanuska-Susitna Valley became one of the most attractive areas for agricultural development in Alaska. Farms and homesteads sprouted up in the region and thrived for a few years, but most of them were abandoned after World War I. It was not until the Great Depression, when a new era was born in Alaska agriculture with the creation of the Matanuska Valley colony, that the region became Alaska's major farm belt.

In 1934-35 the Federal Emergency Relief Administration, one of the many New Deal relief agencies created during Franklin Roosevelt's first year in office, planned an agricultural colony in Alaska to utilize the great farming potential of the Matanuska-Susitna valleys, and to get some American farm families off the dole.

Social workers picked most of the 203 families for the Matanuska Valley colony from the northern coun-

An Alaska Railroad engine steams past the Hikey farm in the Matanuska Valley in 1918. A huge vegetable garden can be seen to the right. Construction of the railroad provided a big incentive for settlers; by 1917 approximately four hundred people lived in the valley. (Lulu Fairbanks Collection, University of Alaska, Fairbanks)

In 1935, two hundred families from the northern Midwest traveled by train and ship to settle and farm the Matanuska Valley. The much-publicized, government-sponsored Matanuska Valley Colonization Project was designed to get and keep families off relief during the Great Depression. The project also aided Alaska's economy by producing more local food, lessening dependence on expensive and unreliable transportation of foodstuffs from Outside. Here, a group of settlers from Minnesota boards an army transport ship in San Francisco. (Lulu Fairbanks Collection, University of Alaska, Fairbanks)

ties of Michigan, Wisconsin, and Minnesota because it was believed that the climate in those areas would be similar to that in Alaska, and that the many hardy farmers of Scandinavian descent in those three states would have a natural advantage over other ethnic groups living in the north. The government chose the Matanuska Valley as the site for the agricultural colony because of its favorable climate, good soils, and easy access to Alaskan markets by railroad. Most of the colonists arrived in early summer 1935 at the tent cities of Matanuska and Palmer, where they drew lots for their 40-acre tracts and went to work.

The failure rate of the farmers in the Matanuska colony was high. Some say the lack of success was due to bureaucratic bungling, or the farmers' inexperience with Alaskan conditions, or just overly high expectations of what could have been accomplished. Most observers agree, however, that the tremendous growth which took place in the Anchorage area during World War II saved the colony from being a total failure. The military boom and the huge increase in the population of the Anchorage area provided a ready market for the products of the remaining colonists and other farmers in the Matanuska Valley. For many years now the Matanuska Valley has produced more farm products than all the other agricultural areas of Alaska combined, and today the valley accounts for nearly 60 percent of Alaska's total agricultural production.

By Outside standards, however, the total crop production is still very small. As historian Orlando Miller has explained, when Alaska became a state in 1959, the Matanuska Valley "had the appearance and atmosphere of a farming area, although the cropland and production of the valley could have been equaled in the corner of one county in a midwestern state." Though millions of acres of land suitable for agriculture have been identified in Alaska, less than one-eighth of 1 percent of the potential tillable lands in the state have been under cultivation in recent years.

Many reasons have been given to explain why Alaska has not developed into the northern breadbasket which early Alaskan farm enthusiasts predicted

Matanuska colonists gather in 1935 to draw for 40-acre farm plots. The colonists spent their first summer in Alaska living in the tents pictured here. (Lulu Fairbanks Collection, University of Alaska, Fairbanks)

Above — *Few men can boast of spending their boyhoods on Alaskan farms, as in this early-day Matanuska Valley scene. The largest number of farms counted in the state was in 1939, when 623 farms were tallied. Although the number of farms today is considerably less than that, productivity has increased greatly. (Lulu Fairbanks Collection, University of Alaska, Fairbanks)*

Right — *Underground pipes from Circle Hot Springs near the Yukon River, northeast of Fairbanks, direct soil-warming water beneath this productive vegetable garden. In this old photograph, steam rises from the spring as it flows through the center of the garden. Today, using underground pipes, Alaska scientists are experimenting with warming garden soil using waste heat from electrical generating plants. (Bunnell Collection, University of Alaska, Fairbanks)*

it would, but a major factor has been economics. As one student of Alaskan agriculture put it, "there has been a tendency to confuse physical possibility and economic reality." In the past, Alaskans mostly harvested crops that they would otherwise be forced to do without. It was far too expensive to ship hundreds of tons of horse feed north, so hay and grains were a major product of pioneer Alaskan farmers. Before the age of refrigeration and air freight, fresh milk, eggs, and green vegetables were impossible to ship long distances, and the local Alaskan farmer was the only available source of such products.

Today transportation is much quicker and more economical than it was during the gold rush. Also, American agribusiness is far more productive than the small-scale farms of yesterday. But meanwhile the cost of producing crops in Alaska has remained extremely high, and Alaskan farmers can hardly compete with the huge corporations that now dominate U.S. agriculture. Produce from the San Joaquin Valley costs less in Fairbanks than vegetables raised locally in the Tanana Valley, and Alaskans pay more when they buy Matanuska Valley milk or eggs than they pay for dairy products shipped to Alaska from the Pacific Northwest.

A few observers have predicted that Alaskan agriculture will eventually disappear as the 49th state becomes more closely intertwined with the national economy. Many are more optimistic, and predict large future exports to Japan of Alaskan barley and rapeseed, as well as pork, milk, and reindeer. Whatever the fate of large-scale agriculture in Alaska, however, the number of gardeners will most likely continue to increase in both the cities and the Bush as more Alaskans learn how to grow gardens in the long days of the northern summer.

Farming in the Far North

Right — *David Rhode washes a fall crop of carrots and potatoes in the clean water of Kenai Lake. (Helen Rhode)*

Far right — *Gardening in Southeast, with its rainy, cool summers and early fall frosts, can be a challenge. Janice and Jerry Taylor's successful garden in Auke Bay, near Juneau, occupies a small space and is raised to help excess rain water drain off. (Jerry Taylor)*

NORTHERN GARDENING ENVIRONMENTS

Alaska has a reputation for having one of the most hostile climates in the world. Winter can stretch from October through April; at higher elevations, ice and snow can last year-round. In the Interior, winter temperatures often plunge below minus 40°. And by midwinter, daylight is brief or nonexistent.

But Alaska in the summertime is a different world. Although the frost-free season is short (generally lasting from late May to early September), long daylight hours encourage round-the-clock photosynthesis. Summer temperatures in Alaska are mild,

An egg carton holds a dozen giant strawberries, grown by Haines gardener Bob Henderson from a variety developed in the early 1900s by Charlie Anway. (Courtesy of the Sheldon Museum and Cultural Center, Haines)

Lush rain forests and relatively moderate temperatures of Southeast mislead gardeners into thinking that vegetables should grow as luxuriantly as the wild plants. In fact, this is one of the most challenging areas in which to garden. The wetness of the soil inhibits oxygen circulation and soil warmth, and causes leaching, which necessitates frequent fertilization. Nevertheless, successful gardening is possible with proper care. (Betty Sederquist)

ranging from the 60°s and 70°s in Southcentral to the 80°s in the Interior.

Despite conditions generally amenable to gardening, there are significant climatic variations throughout Alaska. For example, permafrost gardening in the Brooks Range requires different techniques and plant varieties from those that would be used by someone gardening in the cold, water-logged soil of a southeastern rain forest.

Following are general climatic descriptions of the major regions of Alaska.

Southeastern Alaska

This belt of rugged islands and a narrow coastal strip contains lush rain forest vegetation that often fools newcomers into thinking vegetable gardening conditions here are ideal: compared with the rest of the state, the temperature never gets very low, the growing season is relatively long, and there is plenty of moisture.

But the challenges here are enormous. Unless specialized techniques are used, gardening can be discouraging. Soils are cold because of their high moisture content, evaporation of moisture from the soil surface further cools the ground, and the rain itself is cold. (Scientists have proved that seeds germinate twice as fast in 60° soil as in soil with a temperature of 50°; and in soil with temperatures below 50°, the plant growth rate drops drastically or is virtually nonexistent. In Southeast, soils reach the 60° range only at their surface.)

In spite of these difficulties, Southeast is the site of many spectacular gardens. Haines, one of the driest areas of Southeast, has supported a significant subsistence agriculture since the turn of the century. The teacup-sized strawberries of Haines are legendary

Carol Janda's garden in Gustavus, in southeastern Alaska, shown in August at its peak, features bountiful crops of lettuce, peas, cabbage, and root crops. Several years of intensive soil-building with seaweed, humus, and lime were required before the poor, sandy soil yielded such results. To the right, zucchini grows under a plastic tent. (Jim Luthy)

Above — *Kodiak Island, one of the first centers of agriculture in Alaska, is the site of many successful gardens. This Kodiak resident has chosen to raise vegetables instead of a lawn in his front yard.* (*Ray Ordorica*)

Right — *A field of head lettuce grows in the Matanuska Valley. Lettuce is the state's second most important vegetable crop. In the background is Pioneer Peak (6,398 feet), a local landmark.* (*Bob Cellers*)

Farmers along East End Road in Homer harvest bumper crops of timothy hay in July 1983. The Kenai Mountains and Grewingk Glacier are visible in the background. In 1982, Kenai Peninsula hay accounted for about 18 percent of all hay produced in the state. (Andrew E. Elko)

among local folks. The Gustavus/Strawberry Point area near Glacier Bay has supported small-scale commercial agriculture for many years.

A few broad valleys in Southeast, where soils are relatively infertile and often poorly drained, have provided local forage for horses and cattle: the Stikine River flats, Point Agassiz near Petersburg, the Taku Flats near Juneau, and the Katlian Flats near Sitka.

Gulf of Alaska Coastal Plain

This region has extensive areas of flat land adjacent to the stormy Gulf of Alaska. Inland, north of these flat lands, rise some of the highest mountains in North America.

Although rugged terrain is not as severe a problem as it is in Southeast, Yakutat residents contend with the same high rainfall and poorly drained soils. The Russians made some unsuccessful efforts at farming in this region during the nineteenth century.

At the west of the coastal plain lie rainy, rugged Prince William Sound and the lower Kenai Peninsula. Although localized subsistence agriculture was prevalent from the 1898 gold rush through the fox farming days of the 1920s, far less gardening now takes place. Gardening, however, is starting to regain popularity as local residents learn of specialized techniques for gardening in wet climates.

Cook Inlet-Susitna Lowland

Most of the land used for agricultural purposes in this area is at an elevation of less than five hundred feet. Some of Alaska's most important present-day commercial agricultural activities take place here; the Matanuska Valley has been an agricultural center for both livestock and crop production since the building of the Alaska Railroad through the region in 1916 and 1917, and more significantly since the arrival of the Matanuska Valley colonists in 1935. The western

Kenai Peninsula, site of many homesteads, has supported limited livestock and crop production since before the middle of the last century.

Kodiak Island has nourished a struggling cattle industry since Russian times.

Alaska's Interior

Hot summers, long daylight, and millions of acres of tillable soil give Alaska's Interior the greatest potential for large-scale commercial agriculture. Although permafrost underlies nearly all of the Interior, it has little detrimental effect on agriculture.

The Copper River Lowland, which is underlain by at least one hundred feet of permafrost within five feet of the surface of the soil, has some potential for agriculture, but high mountains nearby often push cold air downward. Thus the area is prone to summer frosts. Drought is also a problem, as moisture from the rainy Gulf of Alaska is trapped by the Chugach Mountains.

The Tanana-Kuskokwim Lowland, a region of tremendous agricultural potential, contains approximately 8.5 million acres suitable for raising crops. The best agricultural land is situated on river plains and slopes up to fifteen hundred feet elevation.

The Yukon Flats and upstream river valleys contain approximately 2.26 million acres of tillable soils, according to the Soil Conservation Service. These lands are partially subject to spring flooding and some require drainage facilities. This region can produce small grains, cool season vegetables, and many warm season vegetables grown through clear polyethylene. Many of the villages along the Yukon River now support thriving subsistence vegetable gardens.

Above — *A blaze of color marks Joe Holty's poppy farm, in Goldstream Valley near Fairbanks. Holty raises the poppies for seed, which he sells to the state Department of Transportation for planting along Alaska's highways.* (Joe Holty)

Right — *This Nikolai garden, photographed in early July, takes advantage of a southern exposure and reflected sunlight from the south fork of the Kuskokwim River.* (Alissa Crandall)

Florence, Dick, and Julie Collins harvest potatoes on their Lake Minchumina homestead in interior Alaska. After digging, they leave the potatoes on the ground for several hours to dry before storing them. (Miki Collins)

The aerial photo opposite shows the site of a three-acre garden tended by students and staff of Covenant High School in Unalakleet, on Norton Sound. The garden produces three thousand pounds of potatoes each year, as well as radishes, turnips, cabbage, spinach, lettuce, and carrots. At left, residents of the school pitch in and help with the potato harvest. (Both photos by James Hjelm)

Western Alaska

This region's climate is moderated by the Bering Sea. Usually defined as including the Bristol Bay area and the Yukon-Kuskokwim river delta, western Alaska's flat landscape is composed of tundra interspersed with thousands of small lakes and meandering streams. Beautiful family gardens grow on sites protected from cold sea winds at Unalakleet, Akiak, Bethel, Platinum, Aleknagik, Dillingham, and Naknek. Successful commercial vegetable production takes place in Aniak.

West of the Bristol Bay area are the wind- and rain-wracked Alaska Peninsula and Aleutian Islands. Because of extreme winds, cool summer temperatures, and heavy precipitation, gardening is especially difficult here. However, sheep and cattle have been successfully raised in this region for many years.

North of the Yukon-Kuskokwim delta lie the Nome and Kotzebue regions. Cool summer temperatures, short growing seasons, continuous permafrost close to the surface of the soil, and cold prevailing winds from the Bering Sea hamper gardening. Wind protection, be it in the form of a fence of old oil barrels or a polyethylene screen, is essential.

Arctic Coastal Plain

This region, poorly drained because the entire area is underlain by permafrost at least one thousand feet thick, is marshy in summer. A network of ice-wedge polygons covers the land, much of which is flat or gently rolling. During summer, the permafrost thaws six inches to four feet below the surface. Because of poor drainage, thousands of tiny lakes and ponds are scattered over the plain. Best drainage and soils are generally along the rivers. Gardening is extremely difficult due to wet soils and the short growing season. Nevertheless, a gardener craving a few fresh greens can

Above — *Peter McKay harvests turnips and radishes from the community garden at Aniak, in western Alaska about 90 miles northeast of Bethel. For several years, the garden has produced vegetables which have been sold in villages throughout the area.* *(Jill Shepherd, staff)*

Right — *Barney Hansen, a longtime resident of Eagle, on the Yukon River, poses next to his immaculately kept vegetable garden in 1976. The Yukon River drainage in interior Alaska is considered the area of greatest agricultural potential in the state.* *(Sharon Paul, staff)*

Produce from this vegetable garden at Bear Lake Lodge, in the Bristol Bay region of western Alaska, helps feed hungry guests. Some areas in this region, particularly around Dillingham, support fine gardens. (Staff)

Above — *This garden, belonging to the Dearborn family in the Matanuska Valley, makes use of microclimates. The garden is planted on a south-facing slope and trees along the edge of the garden minimize the drying, cooling effect of the area's frequent winds.* (Staff)

Left — *Although outdoor gardening in some remote areas can be tough, greenhouses solve some of the climate problems. This corn in a Nome greenhouse produced ripe ears.* (Maxcine Williams)

successfully raise fast-growing, frost-hardy crops such as lettuce and radishes. Use of cold frames and clear polyethylene increase chances of success.

MICROCLIMATES

Microclimates — localized climate conditions — are critically important to productive gardening. When choosing a garden site, take into account the amount of sunshine an area receives, snow levels or drifts, wind angle and force, and distance from frost pockets.

In most circumstances try to plant a garden on a south-facing slope so it receives maximum exposure to the sun. A south-facing slope that also takes advantage of light reflected off water onto the garden increases the effectiveness of the sunlight (this technique is used frequently in Southeast, where so many people live adjacent to salt water).

Avoid planting a garden in low-lying gullies or depressions, which are subject to frost several weeks later in the spring and earlier in the fall than nearby hillsides. To reduce danger of frost, clear an opening through trees or shrubs at the lowest part of the garden, so that cold, heavy air has an escape route.

Wind protection is important. Wind lowers air temperature, blows away fertile soil, causes soils to cool by evaporation, and dries out and sometimes even breaks delicate plants. A windbreak, such as trees or shrubs, or a fence erected so as not to cast shade on the garden, diminishes this problem.

A south-facing wall provides some protection from northerly winds and also reflects heat onto plants.

The creation of artificial microclimates, such as greenhouses, cold frames, and polyethylene mulch and row coverings, is an important facet of Alaskan gardening.

Permafrost

Permafrost, frozen soil or rock found in nearly all of Alaska except for lower elevations in southeastern Alaska, the Alaska Peninsula, the Aleutians, and warmer areas of southcentral Alaska, is not nearly as devastating to the gardener as it can be for buildings constructed on improperly prepared foundations. Usually only minor modifications in gardening practices are necessary.

In areas of continuous permafrost the ground remains frozen below a level of approximately 30 inches as long as an insulating layer of trees, brush, and moss exists; the upper soil refreezes each winter. Removal of the insulating layer exposes the frozen soil to warm sunlight so the soil begins thawing. On cleared areas permafrost recedes to a greater depth each year. On occasion, up to five years may be required to clear, thaw, and drain permafrost soil so a good garden soil results, according to the Soil Conservation Service. However, in areas where the permafrost extends downward for hundreds or thousands of feet, such as in the Arctic, this melting process continues only until a depth is reached where the sunlight no longer has any warming effect.

Gardeners living in areas of discontinuous permafrost often can select a permafrost-free gardening site by inspecting the vegetation and topography. In Alaska's Interior, for example, flat land occupied by spindly black spruce and heavy moss usually contains permafrost. Aspen and white spruce trees usually are found in drier areas and on south slopes where permafrost is lacking or is at depths of more than three or four feet. Often these sites are hilly and well drained. Gardeners living near an office of the Soil Conservation Service often can obtain information about soil conditions of their land. The service is free.

Permafrost — permanently frozen ground — is photographed here on the North Slope. Although permafrost underlies the soil in much of Alaska, it is not a serious problem for gardeners. Special attention to drainage and soil-warming techniques will result in a successful garden. (Neil Johannsen)

Planning and Growing A Garden

Right — *Growing vegetables does not always have to be an elaborate production. If you are short on time and inclination, you might want simply to plant a few rows of radishes, as in this garden in Tanana along the Yukon River, and let the plants fend for themselves until they are ready to eat.* (Alissa Crandall)

Far right — *Space is at a premium in Bob and Rose Arvidson's garden in Cordova, so they make use of containers. In addition to saving space, the containers permit more intensive cultivation and provide good drainage.* (Rose Arvidson)

PLANNING ON PLANTING

Because of the short Alaska growing season, gardens here — more than anywhere else — must be thought out in advance, because when spring arrives everything happens so quickly there is little time for planning.

The kind of garden you grow depends on how much work you want to do, how much experience you've had with gardening, the amount and type of food you want to produce, and the environment in which you live.

For example, if you live in a remote area of Alaska, you may wish to grow some perishable vegetables difficult to obtain locally, such as lettuce, radishes, or

Some handy garden tools are shown here in the yard of Alf and Edna Kalvick of Skagway. Winter is the time to repair or maintain your equipment. (Bob Coffey)

tomatoes. If you live in an area subject to late spring and early fall frosts, frost-resistant vegetables, such as cabbage, beets, cauliflower, peas, and broccoli, may be more appropriate. Southeast gardeners might prefer shade-tolerant plants such as spinach, leaf lettuce, or swiss chard.

If garden space is extremely limited, avoid planting vegetables that take up a lot of space or give small returns for space used, such as peas, potatoes, vine squash, melons, or corn. Vegetables that produce a lot in a small space include tomatoes, radishes, lettuce, beans, bush squash, turnips, onions, and greens.

The crops that provide the greatest vitamin value for the space used are, in decreasing order: carrots, pototoes, onions, beets, broccoli, cabbage, celery, winter squash, turnips, spinach, snap beans, cauliflower, and tomatoes. Vegetables low in nutrition for the space used include radishes, lettuce, and peas.

Be sure not to crowd plants. Allow enough space for convenient cultivation with available tools. If you're cultivating your garden by hand, you may find shorter rows less monotonous. If, on the other hand, you are using power equipment, such as a rotary tiller or tractor, make the rows long to avoid unnecessary turning at the end of each row. If you use a tractor, plant rows far enough apart so equipment does not damage plant root systems and can pass between the rows even when the crops are close to maturity. Arrange rows so they receive maximum sunlight. Plant tall vegetables, for example, on the north side of the garden plot so they won't shade other plants.

Be sure to rotate crops; don't grow vegetables of the same family in or near the same location two years in a row. This may help to prevent diseases that overwinter in soil. Rotating is also helpful in controlling cutworms and root maggots, which affect members of the cabbage family.

Group crops according to height and put perennials, such as rhubarb and chives, to one side or at the end of the garden, or in another portion of your yard, where they won't interfere with spring and autumn soil preparation.

Beets, carrots, and onions — all small crops — can be planted in rows a foot or less apart if soil fertility and moisture permit. These crops also can be planted in wide rows; this technique produces four to seven times more volume in the same area. Scatter seeds in bands four to six inches wide rather than planting in narrow rows. Carrots, beets, turnips, onions, leaf crops, dill, beans, peas, and rutabagas are well-suited for wide-row planting.

Unless you're extraordinarily fond of swiss chard, leaf lettuce, and parsley, plant only short rows or portions of rows for each kind. Each of these plants can be harvested by occasionally picking a few leaves from the outer edges during most of the season; this induces the plants to produce a continual supply of high-quality produce for a longer period.

In Alf and Edna Kalvick's garden in Skagway, plants are grouped according to height to permit maximum exposure to the sun. (Bob Coffey)

A Timetable for Alaskan Gardeners

Summer is a hectic time for the Alaska gardener; nearly everything happens during a few short months. Since so many garden plants need nearly the whole growing season to mature, it is critical that the garden be planted as soon as spring weather allows. A large number of gardening tasks hit the gardener during a few weeks in spring, so it's best to do the necessary planning and seed purchasing while the snow is still on the ground.

Winter (November through February)

— Check your gardening record for the past season. Note vegetable varieties planted and successes or failures with each. Review insect and disease problems and results of controls used.

— Plan to plant only what you can take care of easily. Use records of past gardens to aid in deciding such things as what to grow, how much to grow, how to lay out your garden, varieties to plant, and how much seed to buy. Also keep in mind personal preferences and how many people you hope to feed with your garden.

— Repair or maintain equipment. Make sure garden tools are clean. Sharpen spades and hoes. Cover with a light coat of oil to protect from rust. Make hotbeds or cold frames if desired.

— Purchase any equipment or supplies you will need, including pesticides. Plan to buy only what you will use. If you buy your seed early, you will be more likely to get exactly the varieties you want. Remember to purchase only varieties suited to your area.

March

— If growing transplants from seed, determine planting date by counting back from the average date of the last killing frost in your area.

— In southeastern Alaska remove protective mulch from perennials as weather permits and before new leaves turn yellow. Wood ash, cold frame sashes, or sheets of plastic spread over the snow hasten melting.

April

— Remove any coarse plant material that would interfere with cultivation. Smaller pieces of plant material may be worked into the soil to increase organic matter.

— Six weeks before last frost, start transplants for eggplant, peppers, and tomatoes.

— Four weeks before last frost plant seeds of cabbage, cauliflower, broccoli, cucumbers, pumpkins, winter squash, and, if early crops are desired, head lettuce and kale.

— Continue scattering ashes over the garden plot to hasten the melting of snow and more quickly warm the soil. Ash application can hasten removal of snow by two weeks.

— If you did not rototill or plow the garden in the fall, do so in spring as soon as soil conditions

The Anderson family's garden at Hope, on the Kenai Peninsula, makes use of two simple plant protectors: sheets of clear polyethylene and glass bottles with bottoms removed. Such protective devices allow plants to be put out as early in the season as possible.
(Helen Rhode)

Mark Hamilton rototills his Anchorage back yard before fertilizing, liming, and planting the area with vegetable seeds and transplants. (Staff)

permit. Don't till the soil until it is fairly dry. The soil is still too wet for tilling if a clump, squeezed tightly in your fist, holds its shape when pressed with your finger. If the clump of soil readily disintegrates upon pressing, the soil should be dry enough to till. If you rototilled or plowed in fall, cultivate or dig the soil to a depth of four to five inches as soon as the soil is dry enough.

— After plowing or rototilling and before planting, apply and incorporate any necessary fertilizers.

— In southcentral Alaska remove protective mulch on any perennials as weather permits and before new leaves turn yellow.

— Check for and treat any damage to fruit trees or shrubs from weather, moose, or rodents.

May

— If you were unable to prepare your garden soil last month, do so now, as soon as the soil is dry enough. In many areas of Alaska, snow doesn't even melt until the beginning of May.

— In interior Alaska, remove protective mulch on any perennials as weather permits and before new leaves turn yellow.

— About three weeks before your area's last expected frost, you can begin hardening off frost-susceptible plants. One way to avoid having to carry your plants inside and outside every day is to move them to cold frames, which you open a little more each day. Or you can make individual plant protectors of glass, plastic, wax paper, or any other easily available materials.

— In middle or late May, depending upon soil conditions and weather, plant frost-resistant vegetables such as parsnips, bulb onions, cabbages, lettuce, mustard, parsley, and some varieties of peas. Other plants that can be seeded directly into the garden up to two weeks before the last killing frost include beets, carrots, swiss chard, potatoes, radishes, spinach, and turnips.

— Transplant hardened-off cabbage, broccoli, cauliflower, and lettuce as soon as the plants can withstand a few degrees of frost.

— As soon as the last killing frost has passed, plant frost-susceptible vegetables. If the weather reports indicate a freezing or near-freezing night after you have already planted them, put protective coverings over the vegetables.

June

— Plant protectors may still be necessary for frost-susceptible plants during the first part of June. Types of plant protectors used vary greatly — examples include A-frames, plastic screens, polyethylene sheeting, plastic cold frames, or open glass or plastic jugs with the bottoms cut out.

— Try to do transplanting on a cloudy day, when light and heat shock are small. Be sure roots make firm contact with the soil. Use a starter solution (a high-phosphorus fertilizer dissolved in water) to promote rapid root growth. Transplants must not be allowed to dry out.

— June is a dry month in many areas of Alaska, so water as needed. If you are not able to

The Arvidsons' Cordova garden combines many gardening techniques worthwhile for Alaska gardeners. Vegetables are grown through clear polyethylene to increase soil warmth, the garden is planted in the sunniest location possible, protective coverings are used for sensitive plants, and raised beds and borders facilitate drainage. (Rose Arvidson)

use prewarmed water on the garden, try to irrigate in the morning so the soil has all day to warm the water. During each watering, the soil should be soaked thoroughly to a depth of six to eight inches.

— Weeding is critical during this period. If you don't keep up with the weeds now, they will overwhelm your vegetables throughout the rest of the summer. Take care of weed control following a rain as soon as the soil is dry enough. Control is most effective when weeds are small and roots are shallow; the roots of crops are thus less susceptible to injury. Remove weed parts to prevent regrowth and try to eliminate weeds before they go to seed.

— Throughout the growing season keep an eye out for diseased plants and plant parts and remove them as soon as you see them. Check for insects and insect damage daily and take corrective action as necessary.

— You can begin harvesting early plantings of radishes, lettuce, and greens during June. Thin seedlings, as needed, early so the unwanted plants don't rob nutrients and moisture from plants you want to leave. Many thinnings are edible.

July

— By the first part of this month you can harvest thinnings of lettuce, beet greens, and turnip greens for salads. Thinning is critical during this period to ensure that the remaining plants have sufficient room to grow. And don't forget to keep up with the weeding.

— Spinach and greens are available in early July. Early cabbage and lettuce mature in July, and cauliflower starts producing heads this month.

— You should be able to start harvesting at least the center heads of your broccoli during July. Everbearing varieties of peas should be producing their first ripe pods during this month. Rhubarb is at its best during June and July; by August, the stalks in warmer areas may start getting stringy. Hill potatoes to avoid greening and to allow room for tuber growth.

— In mid-July, if you live in a rainy area where nutrients are quickly leached from the soil, side-dress your vegetables with additional fertilizer. Use a high-nitrogen fertilizer because it is the nutrient most easily lost to leaching. A nitrogen deficiency causes yellowing of leaves.

— Make continued plantings of lettuce and

Far left — *Perhaps the most effective means of warming Alaska's cold soils is by using clear polyethylene mulch. This method makes it possible for Charlie Brent to raise vegetables as far north as Wiseman, in the Brooks Range. (Nancy Simmerman)*

Left — *Throughout the growing season, check plants daily for signs of disease or insect damage, and take corrective action right away. Root maggot infestation is causing the wilting of these cauliflower plants. (D.P. Bleicher)*

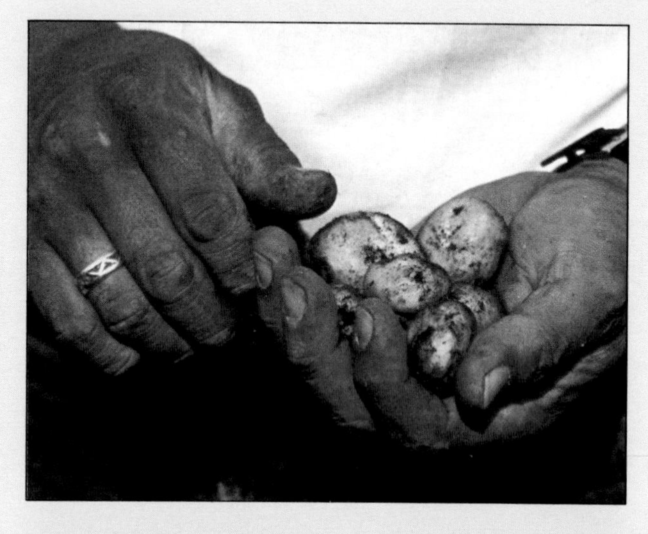

Bob Wilson holds a handful of new potatoes from the Kenaitze Tribe's Wildwood Agricultural Project in Kenai. These tender, tiny potatoes can be dug any time after blossoms (shown below) have formed on the plants.
(Right, Alissa Crandall; below, Bob Cellers)

radishes, if you wish to harvest these crops late in the summer.

— Cutworm season is usually past by this time and the crops are large enough so the few cutworms that remain do not create the same devastation as when the plants were seedlings. If using pesticides, remember to discontinue treatment the required length of time before harvest, according to instructions on the individual labels, and follow all safety precautions carefully.

August

— This is the month when most garden vegetables mature. Harvest the vegetables at the peak of maturity to ensure garden freshness, as well as for best quality if you plan to freeze or can your crop.

— As soon as you've harvested plants that do not produce a sustained yield, pull up the rest of the plant and discard. This practice helps discourage insects and disease.

— Clean, disinfect, and repair the root cellar or storage area, so it is ready for the produce well before harvest.

— In late August, if early frost threatens, cover susceptible crops with blankets, newspapers, or sacking.

— Weed and water as necessary.

September

— Leave root crops that you intend to store (except for radishes) in the ground until hard freezing starts, as this reduces storage time.

Harvest is a busy time for all gardeners — from those with small back yard plots to operators of full-scale commercial vegetable farms. Here, crates of Matanuska Valley potatoes await sorting. (Jeff Schultz)

— When frost threatens, harvest all above-ground crops possible and cover the rest with papers, sacking, or blankets.

— Parsley and other herbs for winter growing indoors can be potted this month and left in a cold frame or cool cellar for a month or two.

— Mulch strawberry and asparagus plants before the ground freezes to protect them against low temperatures. If mulched too early, the plant may continue to grow under the insulating layer instead of going into dormancy.

— Harvest green tomatoes before frost occurs and put them inside, keeping them at a temperature of 50° or more and out of direct sunlight until they turn red; then refrigerate.

— Add garden waste to your compost pile, or till the waste back into the garden soil.

— Before the ground freezes, collect potting soil for next spring's transplants. If you don't do your own soil analysis, collect a soil sample and contact the nearest office of the Cooperative Extension Service to have the sample processed so you'll know the kind and amounts of fertilizer and lime to apply next spring.

— If you are growing bulb onions, pull them up and let them finish drying inside. Spread them in a single layer until the tops are yellow and limp, then cut off the tops, leaving one inch of neck.

Every gardener looks forward to the harvest. Here, the late Vic Mahan proudly displays beets and carrots raised in his carefully tended Seward garden. (Ruth McHenry)

— Dig and store all parsnips and carrots, destroying any that are wormy or otherwise damaged. Store only top-quality produce. Harvest beets, cabbage, turnips, and celery.

— If possible leave potatoes in the ground until frost kills the vines, then wait one week before digging them; this toughens their skins. Digging potatoes with your hands reduces bruising. After digging, keep the potatoes at room temperature for a week to condition them; this gives damaged areas a chance to heal over.

— In areas where wind and water erosion is not severe, rototill or spade the soil after you've pulled up the garden. This practice can provide the opportunity to incorporate stalks, leaves, and other garden waste into the soil. In dry areas of Alaska where it is desirable to enhance garden soil moisture, rototilling, spading, or plowing leaves the soil with a loose, uneven surface that captures and holds the water better than a packed surface.

October

— In Southcentral and Southeast, incorporate lime if needed and plow or rototill after soil temperature has dropped below 50°.

— Harvest celery and brussels sprouts if any plants are still remaining.

— Cut off perennials at ground level.

— In the Interior (where the soil is usually frozen solid by October 1) bend raspberry canes so they will be covered by winter snows. This affords some protection from moose and severe temperatures.

John Bittner, of Watership Farm near Homer, rototills an experimental plot for the raising of annual forage for livestock. (Janet and Robert Klein)

Above — *A worker cultivates with a disk harrow at the experimental farm near Palmer. Proper soil preparation is essential to successful farming or gardening.* (Bob Cellers)

Left — *This rich compost is made up of kitchen scraps, leaves, and garden scraps. It contains no soil.* (Sara Hornberger)

SOIL: THE BASE OF IT ALL

Good soil, defined as a fertile medium for root growth, is one of the most essential aspects of successful gardening. Alaskans are lucky in that regard: although soil scientists generally classify Alaska soils as infertile, they respond magnificently to fertilization.

Proper soil preparation provides for nutrients, soil warmth, balance of acidity/alkalinity, moisture, texture, and oxygen content. The exact type of soil is less important than whether or not it is well drained, well supplied with organic matter, retentive of moisture, and reasonably free of stones. Good garden soil can be created from nearly any type of soil.

Ideally, soil should be a substance that not only holds plants upright but also allows for a free flow of air and water to reach the roots. This porous material should permit excess water to pass through it freely and at the same time retain just enough moisture to enable plant roots to take up nutrients in solution. Essential nutrients should remain continually available to the roots. In addition, soil organisms should be able to carry on the business of breaking down organic matter into a form usable by plants. Good gardening soil should be open-textured and easy to crumble.

Soil Types

Soil types vary with region and topography. Each type requires special treatment. Most soils are not entirely one type but instead are some combination of types. Some basic soil types include the following:

Clay. This type of soil, made up of fine particles invisible to the unaided eye, does not permit water to drain freely. The particles swell when wet and shrink and crack when dry. Most clay soils have lots of plant nutrients, as the nutrients can't leach through. Thus less fertilization is generally required. Clay soils have a lower-than-average oxygen content. Work clay soils when they are moist but not wet; if the soil is too wet, lumps result which become rock-hard when the soil later dries out. Redeem clay soils by adding sand, peat moss, compost, or other soil-conditioning material. Fortunately, clay soils are not widespread in Alaska.

Sand. Sandy soil, which feels gritty when rubbed between the fingers, is made up of tiny pieces of rock. Sand cannot hold water. Nutrients leach out quickly, so large amounts of fertilizer must be applied. Much soil in the Bethel area, for example, is sandy; the soil dries out quickly during the summer. Peat can be added to retain the moisture, although a pint of lime per cubic foot should be added at the same time to counteract the acidity of the peat. Other organic matter also helps to increase the water-holding capacity of sand.

Small rocks generally don't hinder plant growth, unless they are large enough to block the flow of moisture to the plant or decrease the amount of soil available to the root system.

Silt. This type of soil feels something like talcum powder and smears like finger paint. In terms of moisture and nutrient-holding qualities it is halfway between sand and clay.

Organic matter. The most important element of all, organic matter, sometimes called humus, gives soil its spongelike quality of soaking up and storing nutrients and moisture. If your soil appears dark brown or black, it probably has a high percentage of organic matter. Add organic matter to your garden by adding composted material or peat, or incorporating plant debris into the soil. Since humus is so important in the garden, any practice that adds or saves organic matter is important.

Warren Larson, agricultural resource agent for the Cooperative Extension Service in Homer, spreads lime on an experimental plot. In many parts of Alaska, rain and snow leach the soil, making it very acid; applications of lime decrease soil acidity.
(Janet and Robert Klein)

To Lime or Not To Lime

In many areas of Alaska, large amounts of rain and snow leach the soil so it becomes very acid, or "sour." Acidity can also be caused by decay of organic matter. Acidity is measured as pH (positive Hydrogen); a pH of 7 is neutral, levels below 7 indicate acidity, and levels above 7 indicate alkalinity. Silt and loam soils in southeastern Alaska are very acid, with a typical pH of 4.5, so lime must be applied to bring the soils to a level near neutral. Only in some areas of Alaska's Interior that receive little rainfall are soils neutral or slightly alkaline.

Since soil acidity varies by region and since even a well-limed soil may leach and become more acid after a few years, it is important to have your soil tested periodically so you'll know exactly how much lime to apply. Although several kinds of inexpensive soil testing kits are available from local nurseries, you'll get a much more accurate evaluation of your lime needs by having your soil sample tested through the nearest office of the Cooperative Extension Service.

Getting the Soil Ready for the Seeds

When is soil ready for tilling? To find out, use this simple moisture-content test: squeeze a handful of soil into a ball; if the ball sticks tightly together when pushed with your finger, the soil is too wet; if it crumbles easily when pushed, it is ready to be tilled. Soils that are tilled while they are still very wet may later dry into hard, rocklike clods that impede plant growth. Soils in drier areas of Alaska may be ready for tilling slightly earlier in the season than soils in wet areas. In parts of Alaska with soils that are wet year-round, be sure to use proper drainage techniques.

When the garden is ready to be tilled, spade or rototill to a depth of six to eight inches. If you are growing root crops, you may wish to work the soil slightly deeper. About midway through the tilling process, add fertilizer and humus as needed, incor-

Bethel gardeners must contend with permafrost and soil that is 98 to 100 percent sand. Despite the extra care necessary, this Bethel resident has managed to grow a fine crop of potatoes. (Pat Barker)

porating it throughout the surface layer of the soil. After tilling, rake the garden plot to lightly pack the soil, which will slow down moisture evaporation. After raking, plant your seeds as soon as possible so the seeds can take advantage of a maximum amount of soil moisture.

Most of the elements known to be needed for plant growth come from the soil. Like people, plants need a carefully balanced diet. Plant food must be adjusted for Alaska's unique climate and soil conditions.

Warming the Soil

Artificial warming of Alaska's cold soil is beneficial — some say critical — to a healthy garden. Soil temperature is as important as air temperature for good plant growth. Numerous techniques have evolved to warm soil easily and economically.

Along the rainy coast of southeastern and southcentral Alaska, soil warming often goes hand in hand with soil drainage. Perhaps the easiest way to simultaneously achieve these goals is by using raised beds. Soil is built into long rows one to two feet high, and two to three feet wide. The sides of the rows usually are vertical and are supported by boards, small logs, telephone poles, railroad ties, or rocks. Soil temperature is increased approximately 4° when the sun hits the sides of boards used in raised beds. Temperatures of these raised beds can be increased

This beautiful garden grows in a climatically marginal area because of wide row planting, a soil warming technique. Each row of plants is built up one foot above ground level and approximately four feet wide.
(Sue Entsminger)

another 10° to 40°, depending on region, by use of clear polyethylene coverings.

Frequent cultivation, practiced primarily to get rid of weeds, helps warm the soil during the growing season by permitting warm air to circulate freely in the upper layers of the soil.

Straw, newspaper, or sawdust mulches are sometimes used in warmer climates for weed control and moisture retention. They should be avoided in Alaska, however, as they prevent sunlight from warming the soil.

If you have a fairly large garden area, ridging may be an effective method of warming your soil. The sides of the ridges should slant in such a fashion that they are at right angles to the sun's rays during the warmest part of the day. Thus maximum use is made of the sun's energy.

Usually a width of five feet is required for each ridge. The higher the ridge is built, the greater the warming effect, but it may be difficult in some areas of the state to keep the upper part of the ridge moist enough for plants to grow. The ridges should have a width on top of at least one foot, as narrow ridges with only one row of seeds drain and dry out too quickly.

The most dramatic soil-warming technique is that of using sheets of clear polyethylene (plastic) mulch. Soil temperatures are increased up to 40° and 50° in interior and southcentral Alaska with this method.

Black plastic, used with success in more southern latitudes to raise soil temperature, is ineffective in Alaska because it prevents the sun from directly warming the soil. Black polyethylene can be used as a weed mulch, however, with cold soil crops such as members of the cabbage family.

Snap beans, considered a marginal crop in most areas of Alaska, are raised through soil-warming clear polyethylene in Bethel.
(Pat Barker)

FEEDING YOUR GARDEN

Plants need to be given the right types and quantities of nutrients at the right times. The Cooperative Extension Service gives a good summary of plant nutritional needs.

"All plants require at least 16 elements for proper development. Of these, carbon, hydrogen, and oxygen are supplied by air and water. These three elements typically account for 95 percent of the plant's total dry weight. The remaining elements must be provided by the soil or growth medium.

"Six of the 13 elements normally provided by soil are required by growing plants in relatively large amounts. These include nitrogen, phosphorus, potassium, calcium, magnesium, and sulfur. Of these, the first 3 are commonly included in 'complete' fertilizers. The remaining 3, though they may be deficient in certain soils, usually are present in adequate quantities." Most horticulturists recommend that fertilizers contain potassium in the form of potassium sulfate rather than potassium chloride. This adds sulfur, which seems to be advantageous in most areas of Alaska. Also important are iron, boron, manganese, copper, zinc, chlorine and molybdenum.

"Complete" fertilizers contain nitrogen, phosphorus, and potassium, listing the percentage of each on the outside of the bag. For example, a bag of 10-20-20 fertilizer contains 10 percent nitrogen (N), 20 percent citrate-soluble phosphorus (in the form of phosphorus oxide or P_2O_5), and 20 percent potassium (in the form of potash or K_20). Sometimes a fertilizer may contain only one of these three substances, such as 34-0-0. Any additional nutrients also must be named on the outside of the bag.

Fertilizers come in liquid or dry form. Liquid fertilizers are appropriate when fast results are needed. Some gardeners and commercial growers practice a technique called "fertigation," in which liquid fertilizers are mixed in with the water supply.

Dry fertilizers are best applied in spring when preparing the soil for planting. Make sure the fertilizer is mixed in thoroughly, or it may burn seeds and seedlings. Plant your garden as soon as possible after you've prepared the soil.

Applications of fertilizer may also be made later in the season, after the plants have achieved some growth. Apply the fertilizer in a furrow several inches from the plant, taking care not to let this side dressing touch the leaves. If this occurs, wash or dust the leaves immediately. One or two side dressings are recommended during the growing season in rainy areas, where moisture quickly leaches out essential nutrients. In dry areas, such as the Interior, a spring application of fertilizer is usually enough to last the entire growing season.

Slow-release fertilizers, which work on the same principle as time-release cold capsules, need be added far less often. Although slow-release fertilizers offer a longer life, they are very expensive and not normally recommended for use in home gardens. Home gardens need a big boost of fertilizer at the beginning of the season, which slow-release fertilizers can't provide unless applied in very large, uneconomical quantities.

Mineral deficiencies can be difficult for the home gardener to diagnose. Plant growth usually slows down before symptoms of nutrient deficiencies become apparent, but proper applications of "complete" fertilizers usually prevent these problems before they occur. You can contact your nearest Cooperative Extension Service office for assistance.

Composting

Composting is nothing more than taking advantage of the natural process of decomposition of organic materials. Take a walk in a birch forest and scratch a few inches below the surface of the leaves. You will find that the leaves have decomposed into a rich, black soil. In effect, what the plants and trees have taken from the soil is being put back into the soil, with the aid of tiny but hungry soil bacteria and fungi. In the cool climate of the North, composting takes almost twice as long as in the Lower 48.

It makes good sense to recycle what has been taken from the soil; composting is an easy way to accomplish this recycling. The result is valuable humus and

Garden wastes put into this soil-making system in the Baltzo garden near Wasilla emerge three years later as rich humus. Compost, made up of kitchen scraps, horse manure, leaves, lawn clippings, wood ashes, and a sprinkling of chemical compost-maker, is separated by year. The first area is fresh, the second decomposing, and the third crumbling and ready for use.
(C. Howard Baltzo)

Organic Soil Additives

Organic soil additives play an important role as a soil builder. Addition of organic matter improves the physical condition of most soils, increases air- and water-holding capacity of sandy soils, improves drainage of heavy soils, reduces soil compaction, and furnishes a continuing supply of nutrients.

It is usually difficult to apply enough organic matter to supply needed nutrients for plants for a full season, unless large quantities are added. Though gardening success is easier with chemical fertilizers, don't overlook organic additives that will help your garden. Following is a list of a few of these additives and their compositions.

Bloodmeal, dried blood, and bone meal. These materials can be used to supply nitrogen and phosphorus, and to hasten plant breakdown in compost heaps.

Grass clippings. Nitrogen-rich clippings can be worked into the soil or into your compost heap. Be sure not to use clippings from lawns that have been treated with herbicides; the chemicals can kill your garden vegetables.

Leaf mold. Leaf mold is a good source of nitrogen. To make your own leaf mold, shred fresh leaves into a container and keep them damp, applying lime to offset the acidity.

Leaves. Leaves are slightly acid and are a source of humus, calcium, magnesium, nitrogen, phosphorus, and potassium. Apply leaves directly to the soil, as leaf mold, or add to your compost heap.

Manure. One of the best soil additives, manure must be used in large quantities to be effective as a fertilizer. Apply composted manure in the spring as soon as it can be worked into the soil. (Dog manure, though widely available, should not be used on gardens because of the possible spread of dog parasites to root vegetables and thus to the eaters of the crop. Similarly, don't use untreated or raw sewage on your garden.)

Peat moss. Peat moss is made up of sphagnum moss in various stages of decomposition and is relatively easy to find in Alaska. Although peat contains few nutrients, it helps to aerate the soil and improve drainage.

Phosphate rock. This substance contains many important soil additives such as phosphorus, and traces of calcium, sodium, iron, magnesium, boron, and iodine.

Sawdust. Like leaves and straw, sawdust improves the condition of soil by aiding aeration and drainage.

Seaweed and kelp. These ocean plants, abundant for coastal residents, are high in potassium. Seaweed also helps to condition the soil.

Wood ash. This common substance contains calcium carbonate (a form of lime), phosphorus, and potash.

improved aeration, better root and water penetration, reduced crusting of the soil surface, and, in sandy soils, improved water-holding capacity. Composting also adds many plant nutrients.

It is not necessary to do your composting in heaps. An equally effective way of returning organic material to the soil is by strip composting. This technique, which can even be used between plants in the middle of a growing season, consists of spading or rototilling small quantities of organic material into the soil. This material, when covered by soil, should decay in a few weeks. Avoid using straw, coarse dried grass, or dried leaves, which take longer to decompose.

WATER CONTROL

Limited irrigation is necessary in most of Alaska except for parts of Southeast and the rainy outer coast from Bristol Bay south. Sandy soils need lots of water; good loamy soils need a moderate amount; and heavy clay soils need the least watering.

Water supplied in small quantities usually is more harmful than helpful. The shallow layer of moist soil causes the plant roots to grow very close to the surface; lack of root growth causes stunted plants. In addition, frequent small waterings result in more evaporation of water from the soil, cooling the soil more than necessary.

Apply water to the soil slowly so the water doesn't run off; avoid heavy flood-type irrigation. Heavy application of cold water cools Alaska's already too-cool soils even more. Water can be warmed by leaving it in a large, clean oil barrel or other container at the edge of the garden; use this sun-warmed water on the garden and immediately refill the barrel for the next watering a few days hence.

Vivian Menaker of Haines adds wood ashes to her compost pile, thereby reducing the acidity of the compost. Two-by-fours that hold the compost permit ventilation and can be removed as the material is used, making the compost more accessible. (Jack VanHoesen)

If you have pressurized water, watering with a sprinkler is effective; the water droplets passing through the air warm up substantially before hitting the soil, minimizing the cold water problem. Water early enough in the day so plant foliage does not remain wet during the night.

Possibly the most effective irrigation techniques are the various trickle irrigation devices available at nursery supply stores and from catalogs. Trickle irrigation is valuable because the water is applied near the plant and thus less water is required. The areas between the rows are not watered, minimizing weed growth in these areas. Trickle irrigation is especially valuable in situations where only a very limited amount of water is available.

Vegetables should receive an even amount of moisture throughout the growing period. Seedlings in particular should not be allowed to dry out. Fruit trees and shrubs, on the other hand, should be given ample moisture early in the growing season, but water should be withheld at the end of the season to harden off the plants for the upcoming winter.

In much of Alaska, gardeners occasionally have a problem with too much water. Waterlogged soils are cold and contain little oxygen, and excess water leaches away important soil and nutrients. Various corrective measures can be taken to make soggy soils tillable, such as terracing, raised beds, ridges, adding sand and peat, or using drainage ditches or pipes.

Water direct from a well can cool Alaska's already cold soil more than necessary, slowing vegetable growth. Here, well water is sun-warmed in a barrel before it is applied to the garden. (Dorothy More)

FROM THE BEGINNING: SEEDS AND TRANSPLANTS

Getting plants off to a good start is critically important, particularly in the North, where the early part of the growing season is crucial to the success or failure of the garden.

Buying Seed

When buying seed, purchase varieties suitable for the area in which the plants will be grown. Vegetables grown in Alaska must be adapted to two unique conditions: cold soils and long daylight. Many Lower 48 vegetable varieties will not germinate in Alaska's cold soils, which at planting time may have a temperature only a few degrees above freezing. Long daylight causes bolting — formation of premature seed stalks — in many plants, decreasing the quality of the edible product. Recent development of nonbolting varieties has solved this problem.

Left — *Water from Circle Hot Springs is used to irrigate this garden, effectively warming the soil. The garden produces vegetables of exceptional size and quality, which are used by a local resort. (Sharon Paul, staff)*

Below — *Sprinklers irrigate a field of potatoes in the Matanuska Valley. Sprinklers are perhaps the most effective method for watering plants: droplets are warmed by the air before hitting the soil; water accumulates slowly and does not run off; and sprinklers cause the least erosion problems. (Bob Cellers)*

The University of Alaska Cooperative Extension Service publishes an annually updated list of vegetable varieties suitable for your area. The list must be revised often to include new varieties especially suited to Alaska's growing conditions.

Once you have purchased seed, store it in a cool, dry place until you are ready to plant it. For best results, use seed that was packaged for the current growing season.

Seeding Directly to the Garden

Direct seeding to the garden is done if the young seedlings are frost-hardy, tolerant of cool temperatures, difficult to transplant, or capable of producing from seed in a short season.

A few days can be gained on Alaska's short growing season by presprouting seeds. Place seeds on damp paper towels and store them at a temperature of 60° to 70°. Keep the paper damp and examine the seeds every morning. As soon as the first signs of sprouting appear, sow the seed. Presprouted seeds are also useful for starting transplants.

Make sure the soil stays moist during germination and while seedlings are very young. Use of clear plastic is very helpful in retaining moisture for the first few days after you have planted seeds. Remove the plastic or make cuts for the plants to grow through as soon

When purchasing seed, it is important to choose varieties suited to Alaska. New varieties are being developed all the time, partially through the efforts of Agricultural Experiment Station horticulturists. Here, snap beans and tomatoes grow at the Palmer station. The varieties were specially developed to produce ripe fruit outdoors during Alaska's short growing season. (Harvey Bowers)

Soon-to-be gardeners search through a nursery greenhouse for transplants to buy. For those who want a garden, but lack the time necessary to start plants from seed, purchasing transplants solves the problem. (Jane Gnass)

Above — *At a Cooperative Extension Service workshop in Bethel, a makeshift seed flat — a box with a plastic liner — is demonstrated. Commercial-type markers show the kinds of seeds planted. Soil is a mixture of sand and tundra peat.* (Pat Barker)

Above — *Potato plants occasionally form seed balls, similar in appearance to green tomatoes. Potato seed balls, as well as all foliage of potato and tomato plants, should be considered poisonous.* (Aline Strutz)

as seedlings have broken the surface, or you may burn them on sunny days.

Transplants

Because of Alaska's short growing season, many vegetables must be grown from transplants if they are to reach maturity before autumn. Some fast-growing vegetables, such as lettuce, can also be grown from transplants to produce an early crop.

When starting seed indoors, place planted containers in individual plastic bags and close with twist ties, or place them on a tray and cover with a large plastic bag or a sheet of plastic wrap. Germinating seedlings should be kept in a warm location. Keep the seeds moist and avoid direct sunlight.

As soon as the first seeds have germinated, place the pots or trays in a warm, sunny window. When the first set of true leaves appears (usually the second set produced by the plant), the seedlings can be thinned out.

When growing seedlings for transplanting outdoors, proper hardening off is crucial. About a week before you plan to transplant the seedlings, set them outside for an hour or two out of direct sun as long as the temperatures are above 50° and the wind is not blowing hard. The next day, lengthen the period of time you leave them outside, until by the end of the week you're leaving the seedlings out all the time. Bring the plants indoors at this time only if there is danger of late frost. This procedure adapts plants to cooler temperatures and outdoor light conditions. Hardened-off plants may acquire a reddish hue, which is natural and nothing to worry about.

If transplants are grown in flats, use a sharp knife to cut the soil in both directions between the rows to form squares large enough to contain individual plants. Make the cuts several days before setting the plants out in the garden. Water the plants thoroughly after cutting.

Transplanting from pots to garden soil is hard on seedlings, often setting back growth one or two weeks and, in some cases, killing the plants if the procedure is done incorrectly.

Do your transplanting in early morning, late evening, or on a cloudy day, which minimizes drying of roots or soil. Make the holes where you'll be setting the seedlings big enough for the root ball and leave some extra space for adding crumbled soil. To remove the plant from the pot, place two fingers over the soil on each side of the plant, invert the pot, and rap the edge of the pot sharply on something hard. Take care not to damage the leaves.

Set the plant in the hole, and for each plant, pour half a cup of starter solution over the roots. Then fill in the hole with crumbled soil and add another cup of starter solution to eliminate air pockets. You can make your own starter solution with one-quarter cup commercial fertilizer, such as 10-52-17 or 8-32-16, per three gallons of water. Let the mixture steep overnight.

An organic starter solution can be made by filling a barrel one-quarter full of barnyard manure (or one-eighth full of poultry manure), filling the container with water and stirring the mixture several times during the next 24 to 48 hours. Pour one pint of this solution around each plant.

MAJOR ALASKA PESTS

Root Maggot

These insect larvae attack nearly every member of the cabbage family, as well as turnips, onions, rutabagas, and radishes. In many cases, turnips and

By late June, transplants set out in this interior Alaska garden three weeks earlier have recovered from transplant shock and are growing well. (Alissa Crandall)

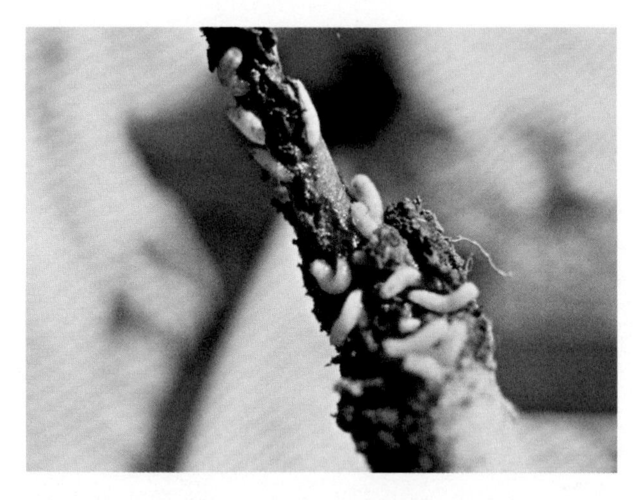

Root maggots, shown above, can cause extensive damage to crops. The turnips below show typical root maggot damage. For home consumption, the damaged portions may be cut away. (Above, Dr. Richard Washburn, USDA; below, David Bleicher, USDA)

rutabagas can be trimmed of damaged portions and are fine for home use.

Maggots are generally yellowish white, legless, and are one-quarter to one-third inch long when full grown.

Root maggots destroy roots and prevent development of plants. They are particularly destructive to young transplants, tunneling along the roots and stems, causing the plants to wilt and die.

Many gardeners agree that this is the only insect pest in Alaska for which chemical control is the only truly effective option. Such control is done for prevention, as the chemicals are effective only from the time the egg hatches until the larva enters the protective cover of the plant root.

Marigolds planted near susceptible plants are reported by some gardeners to be beneficial in controlling root maggots, as the flies are repelled by the scent produced by the flowers. From a practical standpoint, however, the marigolds are too small early in the growing season, when most needed.

New biological controls, such as sex-attractant traps, are being investigated by scientists, but are not yet available on the general market.

Cutworms

These larvae, of which there are many species in Alaska, run in cycles that peak every few years. During years of heavy infestations, cutworms may totally decimate a garden, even attacking plants that are normally cutworm-resistant.

Stout, soft-bodied, and smooth, cutworms may be up to one and one-quarter inches long, and curl tightly when disturbed. They vary in color and may be dull gray, brown, or black, and may be spotted or striped.

The larvae cut off the plants at or above the soil

surface. Some cutworms feed on leaves, buds, or fruits; others feed on the underground portions of the plants. They are particularly destructive when plants are small, as the worms then eat through the whole stems. They are active during the night, and their presence is obvious the next morning when the gardner finds his plants fallen over or missing altogether.

Tilling the soil in September, after the crops have been harvested, may have some value in controlling cutworms, as they pupate at this time of year. In addition, chickens and robins find cutworms very tasty.

Individual plants can be protected by wrapping collars, made from heavy strips of cardboard or brown paper, or old tin cans with bottoms and tops removed, around the base of each plant. Each collar should extend one and one-half inches below the surface of the soil and two inches above it, clearing the stem by half an inch.

If you choose to use a chemical control, apply the pesticide to the soil when the ground is being prepared for planting cole crops (e.g., cabbage, cauliflower, broccoli, turnips, etc.), when damage appears for leaf crops, and as needed for others. Ready-mixed poison baits are effective against species of cutworms that feed above the soil surface.

Cutworms damage crops by cutting off plants at or above the soil surface. The worms typically curl up when handled. (Dr. Richard Washburn, USDA)

Slugs

Slugs, a chronic garden pest in Southeast and along the Gulf of Alaska coast, are also the bane of gardeners throughout southcentral Alaska following mild winters that permit these pests to survive and multiply. They're not easy to control, and if left unchecked, they will attack many garden plants.

The war on slugs is fought by attacking them individually. One method is to place inverted grapefruit rinds or shallow tins filled with stale beer or soda pop, sunk level with the ground, around the garden. Slugs will crawl in and drown in the liquid.

In very rainy climates, trays of beer or soda pop quickly become diluted. Rose Arvidson, a resident of rainy Cordova, suggests laying a bottle half-full of stale beer on its side in the garden; the slugs should be enticed to crawl inside. She also claims that wood ash sprinkled around plants discourages slugs, although rain decreases the effectiveness of the ash.

Helen White, an Anchorage gardener, recommends a ring of sand two to three inches in diameter around

Pesticides

Aside from kids and dogs who go barging through vegetable patches at the wrong times, birds or humans stealing those ripe strawberries or raspberries, and a few very persistent species of weeds, Alaska's gardens have fewer pests than those in other states. Many destructive insects seldom venture North, unless they hitch a ride on a greenhouse plant. Some kinds of plants are totally disease-free, while others are subject to a limited range of small-organism problems that can be controlled either by prevention or by organic means. Only in the case of severe or chronic infestation may pesticides be required.

If you are forced to use pesticides, many of which can be dangerous if mishandled, use caution and common sense. Choose the right pesticide for the job, and read the entire label on every package. Purchase only enough to last one season. Follow the directions exactly. If the label contains cautions about avoiding contact with skin, wear protective clothing. Mix and dilute sprays outdoors, and use only the recommended amounts. Don't smoke or eat while handling pesticides and keep *all* pesticides out of reach of children and pets.

With so many cautions and potential problems with chemical pesticides, it may be easier in the long run to garden preventively: buy disease- and insect-resistant varieties when available; use only vigorous, healthy transplants; and pay close attention to sanitation in your garden.

This field shows patterns typical of cutworm infestation. The worms are active during the night, migrating through the garden and girdling tender young plants. (Dr. Richard Washburn, USDA)

each plant. The sand is supposed to coat the slugs' skins and kill them. Flour has the same effect.

If you use commercial slug bait, which comes in meal or pellet form, place a covering over the bait such as half an empty bleach or detergent bottle (cut out small "doorways" for the slugs to enter); this helps keep out rain, birds, and pets.

Weeds

Last but not least on the list of garden pests are weeds. What is a weed? According to the Cooperative Extension Service, "a weed is any plant that is a hazard, nuisance, or causes injury to man, animal, or the desired crop."

The best weed control is prevention. Begin early in the season — an increasing amount of effort must be expended to get rid of the weeds as the season progresses if they are permitted to grow as they please, flower, and produce seed.

Plant your vegetables in straight, marked rows so you know exactly where to cultivate to get rid of newly emerging weeds.

Unfortunately, it is unwise in Alaska to use labor-saving mulches of straw, newspaper, bark, sawdust, or black plastic as a weed control, as these mulches cool the soil considerably and slow plant growth. Because they are very effective in weed control, however, you may wish to use these mulches with plants that love cool soil, such as members of the cabbage family, if you don't mind smaller or later-maturing plants.

By far the best all-around weed control is early-season cultivation. Even if the weeds are not evident, they are probably there anyway. Frequent light cultivation is better than allowing the weeds to become well established before pulling them.

CLIMATE CONTROL OF SMALL SPACES

Climate control of small spaces is not absolutely necessary to successful Alaska gardening, but can be of tremendous — and often easy — advantage in extending the growing season of many plants and for growing certain varieties that prefer a warmer environment than can be found in Alaska.

Controlling the environment may vary from simple clear polyethylene covering to elaborate, automated greenhouses in which plants may be grown in carefully controlled nutrient solutions with special kinds of indoor lighting.

Outdoor Climate Control

Possibly the simplest technique for outdoor climate control is use of row coverings — tentlike structures — made of clear polyethylene to warm the soil. Many warm climate vegetables can be grown outdoors in warmer areas of Alaska using this technique.

Row coverings are especially advantageous as they generally advance Alaska's short growing season by a couple of weeks. Outdoor tomatoes can be set in the field 10 to 14 days earlier than normal if row coverings are used. Lettuce can be planted 2 weeks earlier with the same technique.

Another simple but valuable climate-control structure is a cold frame; a low, boxlike structure without a bottom and with clear polyethylene or glass covering.

Cold frames traditionally have no supplementary heat, using instead the sun. Thus the top of the cold frame should slope to the south to gain maximum benefit from the sun's rays.

Left — *Young squash plants on a Matanuska Valley farm are individually covered with clear plastic to prevent damage by late frosts.* (Jane Gnass)

Below — *Flowers and vegetables share a raised bed in this Fairbanks garden. In the background, tomatoes and squash grow under a tent of clear polyethylene. The plastic walls can be raised or lowered to control temperature and ventilation.* (Dorothy More)

Greenhouses are structures built for optimum climate control for plants. Two types of greenhouses are illustrated here: above is a simple freestanding greenhouse, the most common and effective type. At right is a lean-to greenhouse, attached directly to the side of a Matanuska Valley home. This type of greenhouse is the most convenient, as it allows access during inclement weather, and can be connected to home heat and water systems. (Above, Pat Barker; right, staff)

This large commercial greenhouse in McGrath is heated with waste heat from the city's diesel power plant. Pipes along the left wall hold strawberry plants (below), which grow hydroponically in liquid nutrient solutions. In addition to strawberries, the greenhouse yields tomatoes and cucumbers which are sold in local markets. (Both photos by Jill Shepherd, staff)

Cold frames are used most often in spring and fall. In spring, seedlings can be transferred from inside to harden off in the cold frame. Some hardy plants, such as lettuce, can be planted directly into the cold frame soil for a crop that will mature weeks earlier than outside-planted seeds. Cuttings also root faster in a cold frame.

A hotbed is simply a cold frame with supplementary heat, frequently provided by manure, electricity, steam, or hot water pipes. Most often, however, heat is provided by a heating cable buried in the soil.

Indoor Climate Control

The most readily available controlled environment is your home. If you lack a greenhouse, seedlings for later transplanting can be started indoors. Unfortunately, light is often limited in a house, contributing to spindly seedlings. Also, the insides of homes are usually too dry for growing transplants, so humidity must be provided by laying plastic sheeting over the just-planted seeds. The biggest problem, however, is temperature control, as many plants are most comfortable at temperature ranges different from those preferred by people. Despite the problems, however, many vegetables can be grown successfully indoors.

Greenhouses

Greenhouses are structures built especially for growing plants under optimum climate control. In Alaska, tomatoes and cucumbers are the two most popular greenhouse crops, but more exotic plants such as grapes, oranges, and even bananas have been successfully grown in Alaska. Greenhouses are also used to extend the growing season and obtain multiple crops of outdoor-hardy vegetables such as lettuce and radishes.

Hydroponics

Hydroponics, the science of growing plants without soil in liquid nutrient solutions, is sometimes used in Alaska's commercial greenhouses; many of the locally grown tomatoes and cucumbers are grown hydroponically.

A big advantage of hydroponics is having exact control over growth conditions. Soil has many unknown factors; you can't reproduce exact soil conditions time after time. With hydroponics there is no soil, so exact control is possible; thus, tomatoes of the same quality can be produced by using the same growing conditions. In addition, hydroponically grown plants need no weeding.

Using hydroponic techniques, lettuce can be grown in Alaska in midwinter. This lettuce, which is especially crisp and tender, can even be grown vertically, on a four- or five-foot-high wall, for example, thus using valuable greenhouse space to maximum.

Strawberries are another crop well suited to hydroponics. In some instances, strawberries have been grown in the top part of the greenhouse, growing upside-down from the roof.

Annuals

This colorful sampling from a garden in Seward proves that beautiful vegetables can be raised even in that rainy, rocky-soiled region. (Ruth McHenry)

	VARIETIES	PLANTING	GROWTH REQUIREMENTS	CARE & HARVEST	
BEAN, SNAP Beans are a marginal crop in cooler areas of Alaska. Don't expect success during cool, wet summers or if you live in an area that experiences midseason frosts.	**Provider** produces smooth, straight green beans about 5½ inches long. **Contender** is similar to Provider, but matures slightly later. **Spartan Arrow** and **Honey Gold** are yellow wax beans that can be grown successfully in Alaska.	Plant a few days before the last frost is expected, choosing a sunny site. Soak seed in water for 12 to 15 hours, then plant no more than 1 inch deep, 2 inches apart, in rows 2 feet apart. Germination rate is usually 3 or 4 days in warm or polyethylene-covered soil; 10 days or more in cooler soil.	Beans like slightly acid (pH 5.8 to 6.5) soil, preferable sandy or silty. Soil should be moist but well drained; light applications of fertilizer are needed. In Southcentral or Southeast, beans grow better through clear polyethylene.	Beans are susceptible to frost, so you may wish to cover them at night until all danger of frost is past. Weed control is important. Harvest depends on personal preference as to size (younger beans are more tender), but usually begins in August.	*Snap bean.* *(Maxcine Williams)*
BEET Beets are one of the easier crops to grow in Alaska. Because of the cool growing conditions, beets do not become woody, so even large vegetables are of good quality.	**Cylindra** produces cylindrical, medium-red roots and reddish green tops 14 to 16 inches tall. **Boltardy** is highly resistant to bolting. **Early Wonder** and **Green Top Early Wonder** produce good quality greens. Other varieties suitable for Alaska include **Spring Red** and **Burpee's Golden.**	Seed directly into the ground, ½ inch deep and 1 to 3 inches apart. Beets are frost resistant and can be planted as soon as the ground can be worked in the spring. Transplanting is not recommended; it may cause malformed roots.	Beets do poorly in acid soil, so check pH (preferred levels are 5.8 to 7.0) and add lime as needed. Soil should be fairly deep, well-drained loam; provide adequate moisture throughout the growing season.	Beets must be thinned as soon as the first true leaves appear; a few weeks later thin again leaving plants 2 to 4 inches apart. Beets can be harvested at whatever size you prefer.	*Beets. (Alissa Crandall)*
BROCCOLI Broccoli grows rapidly in Alaska's cool soils, mild climate, and long daylight hours, producing abundant, tender crops. It needs little care and is high in vitamins A and C.	**Green Duke** is one of the best varieties for Alaska, yielding a heavy crop in the short growing season. **Primo** produces a large, central head and heavy side shoots. Other successful hybrids are **Waltham 29, Green Comet,** and **Coaster.**	Broccoli can be seeded directly into the garden a few days before the last frost, or can be grown from transplants for an earlier crop. Seeds should be planted ½ inch deep and thinned as soon as the first true leaves appear. Transplant to the garden, after hardening off, spacing plants 15 to 24 inches apart in rows 3 feet apart.	Although soil requirements are not critical, broccoli needs plenty of moisture and fertilizer for top-quality heads. The plant prefers a pH of 6.0 to 6.5; use a good general application of fertilizer.	Broccoli is susceptible to cutworms and maggots. Harvest the central head of the broccoli while it is still tight and green. After cutting the central head close to the tip, you can continue harvesting side heads for the rest of the season.	*Broccoli.* *(Penny Rennick, staff)*

		VARIETIES	PLANTING	GROWTH REQUIREMENTS	CARE & HARVEST
	BRUSSELS SPROUTS Growing brussels sprouts is an exercise in patience, as they are very slow to mature. The slow growth is offset, however, by their tolerance of cold — brussels sprouts can be harvested well into the end of the season.	The best variety for Alaska is a hybrid, **Jade Cross,** which produces an abundance of high-quality dark-green sprouts on vigorous plants. **Catskill,** a dwarf variety, is also suited to Alaska gardens.	Because they grow so slowly, brussels sprouts must be grown from transplants. Start seeds 8 to 10 weeks before setting outdoors. Transplant when danger of frost is past, spacing plants 18 inches apart in rows 3 feet apart.	Brussels sprouts thrive in a rich, deep silt loam with ample moisture but good drainage both at the surface and around the roots. Use a general application fertilizer.	Brussels sprouts mature from the bottom of the plant upward. In mid-August, you may want to pinch off the developing tops of the plants, thus diverting the plants' energy into the sprouts. Brussels sprouts tolerate frost well — some even say the flavor is enhanced by frost.
 *Cabbage. (Jack VanHoesen)*	**CABBAGE** This cole crop thrives in Alaska's cool climate and is frost resistant. Some varieties mature even if maximum temperatures never exceed 50°.	**Copenhagen Market** matures early and produces well-rounded, solid heads. **Alaska 6467** is a late-heading variety with a storage life of one year under ideal conditions. **Early Marvel** is one of the best early varieties. Other types suited to Alaska are **Stonehead Hybrid, Early Red Meteor,** and **Hybrid Golden Acre.** For giant cabbages, try **O-S Cross.**	Cabbage can be grown from seed or transplants. Plant seeds ¼ to ½ inch deep and 12 to 16 inches apart. Spacing is important as it controls the size of the heads. Properly hardened cabbage transplants can withstand light frost. Space transplants 12 to 18 inches apart in rows 2 to 3 feet apart.	Cabbage will grow in a wide range of soils, but does best in well-drained sandy and silt loams with a pH of about 6.5. Adequate moisture throughout the growing season is essential.	Watch for pests — cabbage is susceptible to cutworms and root maggots. Heads will crack or split if soil is allowed to dry out, or if drying out is followed by heavy watering. Harvest cabbage when the glossy appearance on top of the head disappears.
 *Carrots. (Miki Collins)*	**CARROT** Despite their slow growth, carrots are one of the best crops for Alaska gardens. They are not bothered by insects or diseases, and are high in vitamin A.	**Super Nantes** has a nearly indistinct core, is of excellent quality, and produces a fair yield. **Royal Chantenay** has slightly tapered, bright orange roots with crisp-textured flesh and fine flavor. Other recommended varieties include **Special Nantes 616, Spartan Bonus, Spartan Sweet, Gold King,** and **Nantes Coreless.**	Plant seeds in early spring as soon as the ground can be worked. Sow seeds ¼ to ½ inch deep, about 1 inch apart in rows 18 inches apart. Thin seedlings early to at least 1 inch apart.	Till to a depth of 8 inches or so to prevent branched roots. Soil should be free of clods and larger stones. Carrots prefer sandy or loamy soil that is slightly acid (pH 6.0 to 6.5). Good soil moisture is necessary for the small seeds to germinate.	Hoe rows to keep down the weeds before the carrots are up. Harvest by working the soil with a spading fork until it is loose enough to pull up the carrots easily.

	VARIETIES	PLANTING	GROWTH REQUIREMENTS	CARE & HARVEST	
CAULIFLOWER Cauliflower is relatively easy to grow, and can withstand light frosts.	**Whitehorse,** best of the early varieties, is usually grown from transplants. **Super Snowball A,** also an early variety, can be seeded directly to the garden and produces compact, white, 6-inch heads. Other varieties include **Snow Crown, Perfected Snowball, Snowmound,** and **Igloo.**	In some areas of Alaska, cauliflower may be direct-seeded for a late crop; however, the plant is usually grown from transplants. After danger of frost is past, harden off and set transplants 16 to 18 inches apart in rows 2 feet apart.	Cauliflower needs very fertile, well-watered soil with a pH of 6.0 to 7.0. Deficiencies of trace elements can make cauliflower difficult to grow in some areas; it is best to contact your local Cooperative Extension Service office for information on these deficiencies and how to correct them.	It is critical to protect cauliflower from root maggots and cutworms. To achieve white heads and best quality, cauliflower must be protected from light. If heavy foliage around the heads does not afford enough protection, draw the leaves together over the head and tie them.	*Cauliflower. (Alissa Crandall)*
CELERY Once celery has passed the critical early growth stages, it is an easy crop to grow in Alaska. Because it takes 10 to 12 weeks to produce a plant large enough to transplant, some gardeners prefer to purchase seedlings from local nurseries rather than trying to grow their own.	**Utah 52-70** is by far the most dependable variety for Alaska gardens. It reaches heights of 28 to 30 inches, producing smooth, waxy, medium green stalks at least 12 inches long. **Utah Jumbo** can also be grown in Alaska.	Start seeds indoors about 10 to 12 weeks before planting date, or purchase seedlings from a nursery. Harden plants off by reducing moisture rather than temperature. Transplant to the garden 1 foot apart in rows 2 feet apart. Keep soil moist.	Because of its shallow root system, celery requires heavy moisture and lots of fertilizer. It prefers muck soils, but will grow in sand or silt. Preferred pH is 6.0, but celery will tolerate pHs between 5.2 and 7.5.	Celery is generally disease free in Alaska. Home gardeners can begin harvesting the outer stalks of celery by midsummer, permitting the inner stalks to continue growing. Harvest the whole plant just before heavy frost is expected.	*Celery. (Alissa Crandall)*
CORN Sweet corn is a marginal crop in the cooler areas of Alaska, but can be grown with great success in warmer parts of the state.	**Yukon Chief** is considered the variety best adapted to cold soils — it germinates in cold soils and matures early. Stalks are 4 to 5 feet high and produce slightly tapered, well-filled ears, 6 to 8 inches long. **Polar Vee** is similar, but matures slightly later. Other varieties include **Earliking, Golden Miniature,** and **Earlivee.**	Sow two seeds, 1 inch deep, at 1-foot intervals in rows 3 to 4 feet apart. At least three rows should be planted side by side for efficient pollination. Growing corn from transplants is generally unsuccessful.	Corn prefers fertile, well-drained soil with a pH of 5.8 to 7.0. Soil warmth is critical to successful corn growing; sheets of clear polyethylene are the most effective way to warm the soil.	Effective pollination is necessary to produce corn with evenly filled cobs. Long spells of cloudy, rainy weather prevent pollination by the wind, so you may have to pollinate the corn by hand. Harvest corn when the silk at the top of the ear begins to dry or turn brown.	*Corn. (Staff)*

		VARIETIES	PLANTING	GROWTH REQUIREMENTS	CARE & HARVEST
Cucumber. *(Maxcine Williams)*	**CUCUMBER** After tomatoes, cucumbers are Alaska's most important greenhouse crop. Outdoor varieties of cucumber can also be grown through clear polyethylene in Alaska's Interior.	**Liberty**, **Victory**, and **Princess** can be grown outdoors through clear polyethylene. Greenhouse varieties include **Gemini**, **Early Surecrop**, **Challenger**, **Progress Burpless**, and **Tabletreat.** European long cucumber varieties suitable for greenhouses include **Fertila**, **Uniflora D**, **Rocket**, **LaReine**, and **Toska 70.**	Cucumbers can be seeded directly into ridge rows in a heated greenhouse, although in most cases seeds are started indoors and transplanted to the greenhouse or garden about 4 weeks later.	Cucumbers prefer well-drained soil, rich in nutrients, and with a pH of 6.0 to 6.8. Fertilizer and moisture requirements are nearly the same as those for tomatoes. Enhancing soil condition with the addition of organic matter is beneficial.	Conventional cucumbers should be trained on a trellis, fence, or strong string tied to an overhead support. Prune the plant back when it has grown above the support. Greenhouse cucumbers must be pollinated by hand, at least once every three days after flowers have bloomed. Cucumbers should be picked as soon as they reach usable size to encourage the set of new fruit.
Eggplant. *(Jill Shepherd, staff)*	**EGGPLANT** Eggplant can be grown as a greenhouse crop in Alaska, or outdoors through clear polyethylene in Alaska's Interior.	**Dusky**, one of the earliest varieties, is the best one for Alaska. It can be grown either in greenhouses or outdoors. **Black-Beauty** is an excellent greenhouse variety.	Eggplant is a slow grower so should be started at least 6 weeks before transplanting. Plants are especially sensitive to transplant shock. At no time during growth should eggplant be permitted to become chilled; even exposure to 50° temperatures will cause the plant to become permanently stunted.	Soil must be warm, well drained, and rich. Preferred pH is between 6.0 and 7.0. Soil moisture should be kept high, and applied at an even rate to prevent blossom end rot.	Pick all fruit before maturity; if the fruit is not kept picked the plants will stop producing. Greenhouse plants are particularly susceptible to damage by aphids.
Right — *Flowering kale.* *(Norma Dudiak)*	**KALE** Kale, a member of the cabbage family, does well in Alaska and is a good source of vitamins A and C. Flowering green or purple kale, often used as an ornamental plant, is also edible.	Although nearly all varieties of kale do well, suggested is **Dwarf Green Curled**, which produces plants 12 to 15 inches tall with a spread of 20 to 25 inches. Leaves are finely curled with a deep yellow-green color.	Although kale can be grown from transplants for an early crop, it also can be direct-seeded into the garden. Plant seeds ¼ to ½ inch deep, 1 inch apart in rows 2 feet apart. Thin to a single plant every 10 to 12 inches.	General soil requirements are similar to cabbage. Adequate fertilizer and water promote fast, tender growth. Kale likes lots of calcium and a pH between 6.0 and 8.0.	Kale is susceptible to cutworms. The plant grows all summer and can be harvested in fall after the first frost, or you can pick the outside leaves as soon as they are large enough. Many people feel that a light frost enhances the quality of kale; even after a moderate frost, it maintains quality and can be harvested in midwinter.

		VARIETIES	PLANTING	GROWTH REQUIREMENTS	CARE & HARVEST
Kohlrabi. (Alissa Crandall)	**KOHLRABI** This plant is easily grown in Alaska gardens. It is raised for the turnip-shaped portion of its stem, which rests on the surface of the ground. The creamy white flesh is high in vitamin C and can be eaten raw in salads or cooked as a vegetable.	**Early White Vienna** and **Early Purple Vienna** are old standbys in Alaska gardens, but many new varieties are being developed.	Sow seeds directly into the garden ½ inch deep and 1 to 2 inches apart, in rows 18 inches apart. Thin to one plant every 6 to 8 inches. For an earlier crop, grow from transplants.	Like cabbage, kohlrabi thrives in any good fertile garden soil. Preferred pH is between 6.0 and 7.0	Kohlrabi grows rapidly and withstands frost. The stem should be harvested when it reaches 2 to 3 inches in diameter.
Leaf lettuce. (Ted Bell)	**LETTUCE** Lettuce is one of the easiest crops to grow in Alaska; it grows so rapidly that it is extraordinarily tender and sweet.	Lettuce is grouped into four types: cos (romaine), butterhead, leaf, and crisphead. Suggested cos varieties include **Little Gem, Paris White,** and **Dark Green Cos. Buttercrunch** is the best butterhead variety for Alaska. Proven leaf lettuce varieties include **Salad Bowl, Grand Rapids,** and **Ruby.** Recommended crisphead varieties are **Great Lakes 659, Minilake, Oswego,** and **Ithaca.**	Lettuce can be seeded directly to the garden, or grown from transplants for an extra-early crop. Lettuce is hardy and can be seeded up to 2 weeks before the last frost; plant seed ¼ to ½ inch deep, ½ inch apart, in rows 18 inches apart. Thin plants to 6 to 12 inches apart, depending on type. The thinnings are excellent for salads.	Lettuce prefers a well-drained, cool, slightly acid (pH 6.0 to 7.0) soil. Liberal use of fertilizer will ensure rapid growth.	Cutworms and slugs are a problem, as is tip burn (caused by nutrient imbalances or uneven moisture, heat, or cold). Lettuce has some frost resistance and can be harvested throughout the growing season.
Melons. (Staff)	**MELON** Melons are a marginal crop in Alaska unless they are grown in greenhouses, a rare practice as the plants take up so much room. A few varieties can be grown outdoors in the Interior through clear polyethylene, but all melons are very sensitive to frost.	**Yellow Baby** and **Sugar Delicata,** watermelon varieties, have both been grown successfully through clear polyethylene in Fairbanks. Two varieties of muskmelon, **Mainerock** and **Alaska,** have also proven to be successful.	Grow melons only from transplants. Plant seed 4 to 5 weeks before setting into the garden or greenhouse. Good temperature and moisture control are essential during early growth stages. Transplant to the garden only after all danger of frost is past.	Good, well-drained, well-fertilized sandy and silt loam is needed for growing melons. Moisture levels should be consistent, and pH should be slightly acid to neutral.	Watermelon is ready to harvest when a rap on the fruit creates a dull sound. Muskmelon should be picked at what is called the "slip" stage — when a slight pressure at the point where the stem joins the melon causes the melon to slip off the vine. Melons must be harvested before first frost.

	VARIETIES	PLANTING	GROWTH REQUIREMENTS	CARE & HARVEST	
ONIONS Onions grow satisfactorily in Alaska. Because of the long daylight, bulb onions do not become very large. Garlic, green onions, leeks, and shallots perform similarly. Chives, however, grow very well because they are winter hardy.	Bulb onions are best grown from sets; no specific variety is needed for Alaska. Suggested green onion varieties include **Bunching, Evergreen White, Beltsville Bunching** and **Evergreen Bunching.**	Direct-seed green onions ¼ to ½ inch deep and 1 inch apart. Start transplants 8 to 10 weeks before planting. Onions are somewhat frost resistant when young and transplant easily.	Onions need a warm, well-drained, well-prepared soil with a pH of between 6.0 and 7.0. Their limited root systems respond well to a general application of fertilizer.	Bulb onion stems can be trimmed when they reach a height of 4 to 6 inches (the trimmings are edible as salad greens). Be sure to thin green onions early. Bulb onions can be harvested throughout the summer, even if the bulbs are small; green onions can be picked as soon as they are large enough to use.	*Onion. (Bob Overmon)*
PARSLEY This herb thrives during cool Alaska summers and has no apparent insect pests or diseases. Since parsley seed germinates slowly, many Alaska gardeners buy seedlings from nurseries.	Most varieties grow well in Alaska. **Extra-Curled Dwarf** is one suggested variety. Another is **Double Curled,** which is frost resistant, early, and rapid growing with dark green, finely curled leaves.	Start seed indoors at least 6 weeks prior to outside transplant date. Plant seed ⅛ to ¼ inch deep, ½ to 1 inch apart. Seeds sprout better if soaked in water first for 6 to 12 hours. Seeds germinate 12 to 21 days after planting.	Parsley will grow in all types of fertile soil and has no special requirements.	No special care is required. Once plants have attained a height of approximately 10 inches, sprigs of parsley can be harvested for the rest of the season until a severe frost occurs. Parsley can be dried or frozen for future use, or several plants can be placed in pots for indoor growth during the winter.	*Parsley. (Alissa Crandall)*
PARSNIP This delicately flavored root crop requires a long growing season and is the last crop to be harvested in fall.	**Hollow Crown,** the most suitable parsnip variety for Alaska, has uniformly tapered roots, 8 inches long with shoulders (root tops) 2 to 3 inches wide.	There is little danger of loss from frost even if parsnips are seeded up to 2 weeks before the last killing frost. Grown like carrots, parsnips emerge and grow slowly. Sow seeds thickly, ½ inch deep, 1 inch apart. Thin plants to 3 or 4 inches apart in rows 2 feet apart.	Prepare the soil to a depth of at least 8 inches. Rich soil, with plenty of organic matter and pH of between 6.0 and 7.0, is needed.	Delay harvest at least until after a sharp frost. Roots may be safely left in the ground over the winter, protected by mulch, and used in the spring before growth starts. Flavor is said to be improved by freezing weather.	

	VARIETIES	PLANTING	GROWTH REQUIREMENTS	CARE & HARVEST	
PEA This plant thrives in Alaska's cool, moist weather and is a dependable, easy crop to grow. Peas are one of the highest protein-yielding crops per acre that can be grown in northern latitudes.	**Green Arrow** is the best determinate (suited to freezing and commercial use) variety for Alaska. Others include **Early Frosty, Signet, Perfected Freezer,** and **Hyalite. Freezonian** is the best indeterminate variety for Alaska. Edible podded pea varieties include **Dwarf Gray Sugar, Sugarsnap,** and **Little Sweetie.**	Peas may be the first seeds planted in the garden in spring as soon as the soil can be worked. Plant seeds ½ inch deep, 10 seeds to the foot.	Peas grown in Alaska require the same amount of nitrogen as other garden vegetables, so apply a general application garden fertilizer as needed. Soil pH should be slightly acid, between 6.0 and 7.0.	Peas thrive in moist weather and are not usually bothered by insects or disease, except for cutworms. Remove pods just before the peas reach full size, but when they are firm and well filled.	*Peas. (Ted Bell)*
PEPPER This frost-sensitive plant is normally grown as a greenhouse crop in Alaska, but can be grown through clear polyethylene in warmer parts of the state. Peppers also adapt well to pot or container gardening.	**Bell Boy** and **Pennwonder** are greenhouse bell peppers of excellent quality and high production. **New Ace Hybrid, Canape,** and **All Big** can be grown either in the greenhouse or outdoors through clear polyethylene. Varieties of greenhouse hot pepper are **Hungarian Yellow Wax** and **Anaheim.**	Plant seeds in pots 6 to 8 weeks before transplant date, at a depth of about ½ inch. Plant 2 or 3 seeds per pot, later thinning to 1 plant. Seedlings emerge in 10 to 14 days. Ideal temperatures are 65° to 70°.	Peppers prefer a well-drained, moist, warm soil well supplied with humus. In a greenhouse, slightly less fertilizer is required than for tomatoes or cucumbers.	Warm temperatures are essential to pepper growth. Nighttime temperatures of 55° to 60° can cause blossom drop. Peppers grown in greenhouses are very susceptible to aphids. Peppers are ready to harvest when the desired fruit size is obtained and the fruit is firm and thick-walled.	**Left** — *Peppers. (Staff)*
POTATO Today, potatoes rate as Alaska's most important commercial vegetable crop. Although they require somewhat more care than other garden standbys, potatoes thrive in Alaska's cool, acid soil.	**Alaska Frostless** is resistant to early frosts and is of high quality but low yield. **Alaska Russet** is a white russet excellent for baking. Other varieties suited to Alaska include **Snowchip, Stately, Kennebec, Ontario, Red Beauty,** and **Bakeking.**	Try to purchase only Alaska certified seed. Don't plant grocery store potatoes; they may have diseases, or may have been treated with a sprout-retardant. Plant as soon as the soil can be worked. Plant each potato section (each with 1 or 2 eyes or sprouts), 1 to 2 inches deep, covering with enough soil to leave a mound, 10 to 12 inches apart in rows at least 3 feet apart.	Potatoes prefer an acid soil and will tolerate a pH of 4.8 to 6.5. Before planting, apply a band of commercial fertilizer along the bottom of the furrow, covering with at least 2 inches of soil.	When shoots are 9 inches tall begin hilling potatoes by drawing up the soil around the newly formed plants to cover the tubers. Potatoes can be dug anytime after blossoms have formed. Potatoes can be left in the ground several weeks after a killing frost, provided the ground does not freeze.	*Potatoes. (Staff)*

	VARIETIES	PLANTING	GROWTH REQUIREMENTS	CARE & HARVEST	
PUMPKIN Pumpkins are considered a marginal crop in most of Alaska, as they must be grown through clear polyethylene to achieve any degree of success. Because of the large amount of room pumpkin plants take up, they are impractical as a greenhouse crop.	**Cheyenne Bush** is a proven variety in southcentral Alaska. **Connecticut Field** is the variety marketed commercially in the Interior. Two other varieties grown in the Interior are **Jackpot** and **Spirit.**	Sow seeds, 5 weeks before transplanting to the garden, 1 to 2 inches deep, 3 seeds per pot. Thin later to the single most vigorous plant in each pot. Don't transplant outdoors until all danger of frost is gone. Transplant to hills 3 or 4 feet apart in rows 6 to 8 feet apart.	Pumpkins require high-quality fertile garden soil with a general application of fertilizer. Soil should be slightly acid, with a pH ranging between 6.0 and 7.0	Harvest pumpkins before a heavy frost, or when they turn a good, mature orange color. Unripe pumpkins occasionally ripen off the vine if they have already started to turn yellow when picked.	**Left** — *Pumpkin. (Staff)*
RADISH Radishes are a good crop for beginning gardeners as they grow rapidly, produce high yields, and are not too fussy about soil conditions.	Among the red radishes, **Champion** is suitable for both greenhouse and home gardens. **Cherry Belle** is an early red-stemmed radish with crisp, white, firm flesh. White radish varieties suggested for Alaska include **Giant White Globe, Burpee White, White Icicle,** and **French Breakfast.**	Plant seeds about ¼ inch deep, in rows 1 foot apart. Thin seedlings to 1½ inches apart. Plant a few feet at a time every 10 days for harvest throughout the summer. Seeds germinate in 4 to 7 days.	Soil should be slightly acid, with a pH between 5.5 and 7.0. Be sure to supply adequate moisture; lack of moisture in the soil causes hot flavor and woody texture.	Because they are fast growing, radish crops can be harvested 3 to 6 weeks after planting. They are susceptible to root maggots, however, so be on the lookout for these pests, especially in later plantings.	*Radishes. (Helen Rhode)*
RUTABAGA Closely related to turnips, rutabagas are planted and grown in the same manner. They are one of the easiest vegetables to raise in Alaska gardens.	**American Purple Top, Golden Neckless,** and **York** are all suggested varieties for Alaska.	Sow seed directly into the garden as early as 2 weeks before the last frost, ¼ inch deep, in rows 20 inches apart. Thin seedlings to 8 to 10 inches apart when they are established.	Rutabagas do best in deep, rich, sandy-loam soil which has been well limed.	Rutabagas are susceptible to root maggots. Harvest rutabagas after the first exposure to frost but before a heavy freeze. Pull, clean, and wax rutabagas for maximum winter storage. Quality can be maintained longer with cool, moist storage, preferably just above freezing.	

	VARIETIES	PLANTING	GROWTH REQUIREMENTS	CARE & HARVEST
SPINACH Variety is crucial to successful growing of spinach in Alaska. Because of the long daylight hours, most varieties will develop a seed stalk before the plant is large enough to eat.	**Melody Hybrid** is a recently developed variety that produces beautiful, big plants with nicely crumpled leaves. Another suitable variety is **Marathon Hybrid,** a vigorous, early plant which produces high yields.	Direct seed to the garden 2 weeks prior to the last killing frost. Plant the seed ¼ inch deep, 2 to 4 inches apart in rows 2 feet apart. Thin seedlings to 8 to 10 inches apart.	Acidity is important to spinach. It grows well only in a pH range of 6.0 to 7.0. Spinach should be planted in well-drained fertile loam, rich in humus and nitrogen.	Harvest the outer leaves of the plant gradually, or the whole plant all at once.
Summer squash (zucchini). (Nancy Simmerman) **SQUASH, SUMMER** Plant just a few innocuous summer squash and stand back! By midseason, under good growing conditions, you'll have more squash than you'll know what to do with. Also, because of their rapid growth rate, Alaska zucchinis remain tender even if they are allowed to grow to a large size.	Both zucchini and crookneck squash can be grown in Alaska. Suggested zucchinis are **Storrs Green Hybrid, Black Zucchini, Caserta, Aristrocrat, Elite, Apollo,** and **Blackjack. Dixie Hybrid Yellow Crookneck** is a new variety which yields an early crop of bright yellow squash.	Begin seedlings indoors in cooler areas, 4 weeks before transplant date, sowing 3 seeds, 1 to 2 inches deep, per pot. Thin later to the most vigorous plant in each pot. In warmer areas, direct seed 7 to 10 days before the last spring frost. Space plants 4 feet apart in rows 5 feet apart.	Summer squash needs well-drained soil and moderate fertilization. Soil should be rich, sandy, and have a pH between 6.0 and 7.0.	Summer squash is ideally harvested after it has reached a length of 4 to 6 inches for crookneck and about 8 inches for zucchini, but it is still tender at much larger sizes. If you can easily puncture the skin of the squash with your fingernail, it is still tender enough to use.
Winter squash. (Ted Bell) **SQUASH, WINTER** Only a few varieties of winter squash are suited to Alaska, as they require a long growing season. The squash, however, will keep for several months under good storage conditions.	**Golden Nugget, Faribo Hybrid R, Kindred** and **Golden Hubbard** yield 4- to 5-pound fruits with orange skins. **Golden Nugget** is a bush type, and **Buttercup** is a dark green squash.	Start transplants 4 to 5 weeks prior to setting outdoors. Plant 3 seeds, at a depth of about 1 inch, per pot. As soon as seeds have formed true leaves, thin to the most vigorous seedlings in each pot. Transplant outdoors when all danger of frost has passed.	Growth conditions are the same as for summer squash.	Grow winter squash through clear polyethylene for best results. Leave squash on the vine as late in the season as possible to allow maximum maturity. Harvest just before the chance of hard frost by cutting the stem 1 to 2 inches from the fruit. Maturity is important — orange squash is mature when it has turned from light yellow to dull orange; green squash is mature when it loses its glossy appearance and becomes dull gray green.

	VARIETIES	PLANTING	GROWTH REQUIREMENTS	CARE & HARVEST	
TOMATO Traditionally the most widely grown greenhouse crop in Alaska, in recent years tomatoes have also become an outdoor crop. However, tomatoes can be difficult to grow: they are susceptible to a number of diseases, and require specialized growing conditions.	Varieties suitable for greenhouses include **Tuckcross 520, Fantastic Hybrid, Tropic,** and **Burpee Hybrid.** Suggested outdoor varieties include **Early Tanana** and **Sub Arctic 25.**	Start seed indoors 8 to 10 weeks before transplanting. Sow seed in flats ½ inch deep in rows 2 to 3 inches apart, about 10 seeds per inch, and cover with clear plastic. As soon as there is any evidence of germination, remove the cover and place flats in full sun. Indoor and outdoor plants must be hardened to withstand their new environments before transplanting.	Even moisture is particularly important to successful growth of tomatoes. Warm soil is crucial; if growing tomatoes outdoors, accomplish this by using a clear polyethylene ground covering. Fertilize tomatoes adequately. Preferred pH is 6.5 to 6.8.	Maintain a careful balance of soil nutrients and moisture. Prune tomatoes to a single stem and stake plants to save space. Tomatoes from several of the recommended varieties will ripen indoors if they must be picked green to avoid frost damage.	 *Tomatoes. (Bob Cellers)*
TURNIP This crop is relatively easy to grow in Alaska, but is subject to root maggot attack.	**Seven Top** is grown for its greens, but may develop objectionable seed stalks. **Early White Milan** and **Purple Top Strap Leaf** are productive and early. Other varieties include **Tokyo Cross, Petrowski,** and **Purple Top White Globe.**	Seed directly to the garden as soon as the soil can be worked. Sow seed ¼ to ½ inch deep, 2 to 3 inches apart in rows 18 inches apart. Although turnips usually are direct seeded, they often grow faster after being transplanted.	Since rapid growth is essential for good root quality, well-enriched soil is necessary. Preferred pH is 6.0 to 7.0.	Thin turnips before the tap roots become fleshy. Tops from turnip thinnings make excellent greens. Harvest turnips when they have reached a diameter of 2 or 3 inches.	 *Turnip. (Staff)*

Perennials

Clusters of bright-red Rescue apples adorn this branch. An apple-crab apple cross, this variety is very hardy in southcentral Alaska. (Staff)

	VARIETIES	PLANTING	GROWTH REQUIREMENTS	CARE & HARVEST	
APPLE Apple trees of the type that bear full-sized fruit are at best marginally hardy in Alaska, although extensive research has been done to find apple varieties that thrive in the North.	Crab apples can be grown with a fair degree of success in most areas of Alaska. **Siberian** and **Red Siberian** are very hardy. A few standard varieties have been grown in milder areas of the state, including **Summerred, Wealthy, Northern Spy, Rescue,** and **Yellow Transparent.**	Apple trees from nurseries have been subjected to a very long period of dormancy; get them started by placing in a sunny spot until the leaf buds break. Dig a hole 6 inches wider than the roots and plant the tree to the same depth at which it grew in the nursery. Partially fill the hole, tamp soil down, water thoroughly, and let water soak in before filling in hole.	Use a complete, well-balanced fertilizer (similar to those used for vegetables) in the spring only. Don't irrigate after mid-August unless the soil and tree are extremely dry.	Anything that can be done to create warm microclimates helps increase chances of the tree's survival. Do pruning in the spring, just prior to the beginning of active plant growth.	*Apples (Summerred). (Staff)*
ASPARAGUS Asparagus is a marginal crop in Alaska. It is grown most successfully in areas offering a consistent winter insulating snow cover, such as the Tanana Valley.	**Faribo Hybrid** is the best variety for Alaska gardens, producing thick, straight green spears with purplish tips.	In Alaska, asparagus is planted at a relatively shallow level because of cold soils. In spring, as soon as the soil can be worked, dig trenches 18 inches wide and 12 inches deep. Mix generous amounts of organic matter and a complete fertilizer with the removed soil, and refill trenches to within 8 to 10 inches of ground level. Place asparagus crowns 18 inches apart and cover with several inches of soil.	Asparagus produces best in peat soils. The soil should not have a pH of less than 6.0. The plant is a heavy feeder, so maintain soil nutrients with frequent applications of fertilizer.	Harvest asparagus when spears are tender, not too thick, and have compact tips. In the Interior, the first spears are ready for harvest by mid-June. A new bed should not have spears cut until the third year after planting.	

		VARIETIES	PLANTING	GROWTH REQUIREMENTS	CARE & HARVEST
Right — *Currants.* *(Maxcine Williams)*	**CURRANT** A few small groupings of these shrubs, if properly cared for, will annually produce many quarts of tasty red or black berries. Although wild currants abound in Alaska, the domestic varieties are more productive.	**Red Lake** produces large, high-quality light-red berries, excellent for jelly, jam, and pies. Other successful varieties include **Holland Longbunch, Stevens #9,** and **Swedish Black.**	Plant rooted cuttings in early spring, as soon as the soil can be worked. Currants also can be propagated by mound layering. Pull a low branch to the ground, slit the underside of the bark, and weight it down. Heap soil over the branch where it touches the soil, allowing the tip to rise above the soil. At the end of the season or the following spring, enough roots should have formed to permit severing the new bush from the parent plant.	Currants prefer a fertile sandy or clay-loam soil with good drainage and a pH of 6.0 to 7.0. Fertilize in spring; don't fertilize or water heavily late in the season, as this reduces winter hardiness.	Most currants bear on canes 2 to 3 years old. Prune bushes to 3 or 4 canes of each age, removing weaker young and old canes.
	GOOSEBERRY Although gooseberry shrubs grow well in Alaska and produce a delicious berry suitable for jams and jellies, they are not as popular as they might be in Alaska gardens.	**Pixwell** grows in both southeastern and southcentral Alaska and produces fine quality berries that are large, oval, and pink when ripe. **Champion** produces large green berries that ripen late in the season.	Propagate from rooted cuttings or by layering in the same manner as currants. Gooseberries, however, are more reluctant to produce roots from cuttings. When setting out in the yard or garden, space the plants 4 to 5 feet apart in rows 6 feet apart. Plant in spring as soon as the soil can be worked.	Gooseberries prefer cool, heavy soil with a large percentage of humus. Supplement each spring with complete fertilizer spread around the base of the shrubs. Preferable pH is between 6.0 and 7.0.	Gooseberries are very shade-tolerant, but need good air circulation. The plant makes a good windbreak. To pick the berries, which ripen all at once, wear leather gloves to strip the berries from the branches.

	VARIETIES	PLANTING	GROWTH REQUIREMENTS	CARE & HARVEST	
RASPBERRY Raspberries thrive in Alaska and are well worth growing. Although after planting they can more or less take care of themselves, the bushes are far more productive if properly pruned.	**Latham,** which does well in all major growing areas of Alaska, produces a heavy crop of large, bright-red berries. **Chief** is characterized by early and heavy fruit on strong canes. Other successful varieties include **Boyne, Indian Summer,** and **Trent.**	Plant as early as possible. One-year-old stock produces the best roots. Dig holes deep enough to contain the roots without crowding. Set the plants 1 to 2 inches deeper than they grew in the nursery. Water thoroughly after planting, and cut back canes to 4- to 8-inch stubs.	Raspberries will produce in almost any soil, although they prefer a pH of 5.0 to 6.5. Fertile soil produces bumper crops. Apply a good general application commercial fertilizer.	Raspberry canes are biennial; they grow up as suckers from the roots in one season and become woody, bear fruit, and die in the second season. In spring, before plant growth starts, remove dead canes as close to the soil as possible. A plant usually thrives indefinitely if well cared for.	*Raspberries.* *(Kathy Doogan, staff)*
RHUBARB Once established, rhubarb keeps on producing indefinitely. It is often grown as an ornamental in Alaska gardens.	**MacDonald** is a red-fleshed variety that cooks to a medium-red sauce. **Canada Red** is darker red and slightly more productive, but lacks MacDonald's quality. Both of these types can be raised in all major gardening areas of the state. **York** is a variety suitable only to southeastern Alaska.	Rhubarb is generally propagated by division in early spring. Firm the soil well around the newly planted root sections, each of which should have a root bud. A light harvest should be available by the second year, a fairly heavy harvest by the third year, and full harvests every year thereafter. Divide and separate as plantings become thick.	Plant near house foundation for early season production. Rhubarb is a very heavy feeder, so use plenty of general application fertilizer, scattered around each hill every year in early spring.	Rhubarb leaves are so high in oxalic acid that they are poisonous. To harvest the stalks, pull them gently out of the ground rather than cutting them. No more than half of the stalks should be harvested from each plant each season to leave sufficient leaf surface for photosynthesis.	**Left** — *Rhubarb.* *(Helen Rhode)*
STRAWBERRY Hardy varieties of these delectable berries thrive in Alaska. Gardeners have been raising strawberries in the North since before the turn of the century.	Two strawberries developed in Alaska, **Sitka Hybrid** and **Alaska Pioneer,** are very hardy, producing small to medium fruit. Other successful varieties include **Matared, Susitna,** and **Ogallala.**	Spring is a good time to plant strawberries, but young plants can be moved easily all summer. New runner plants don't bear fruit the first season. When planting strawberries, make sure the crown of the plant is even with the surface of the ground and firm the soil around the roots.	To maintain berry production, it is important to provide lots of water and nutrients (use standard commercial garden fertilizer). Strawberries like a pH of 5.5 to 6.5. The plants grow best in fertile sandy or loamy soil, but may be grown in any good garden soil that is well drained and well supplied with organic matter.	During the first summer pinch off all the blossoms on young plants once a week until July to permit the plants to develop strong root systems. Clear polyethylene can be used to enhance strawberry growth in areas that receive relatively little rain.	*Strawberries.* *(Ruth McHenry)*

Grains

Right — *Rocky Goodwin bales hay on his farm near Palmer. The bales will be stored for winter use as livestock feed.* (Bob Cellers)

Far right — *A golden field of weal barley grows at the University of Alaska's Agricultural Experiment Station near Palmer in the Matanuska Valley. The experiment station works to develop new varieties of grain which produce well in Alaska.* (Bob Cellers)

Editor's note: *Much of the information for this chapter was provided by the University of Alaska, Agricultural Experiment Station and Cooperative Extension Service.*

Man has recognized the importance of grain for centuries — that it is a high-energy food for man or animal and is easily stored and transported. Today, grain products make up about 20 to 25 percent of the average diet in the United States. As a group, grains are the most adaptable crop species and can be grown in virtually any environment.

Three major grain types are suited to growing conditions in certain areas of Alaska: barley, oats, and wheat. Known collectively as small or cool-season

Oats grow near Bodenburg Butte in the Matanuska Valley. The valley produces a small amount of grain, most of which is used for hay and silage. (Bob Cellers)

grains, they all have either winter or spring growth habits.

In Alaska, grain is grown primarily for use in livestock feed, and more than 90 percent of the state's grain production takes place in the Tanana Valley. Small amounts of grain (mostly oats) are grown for hay and silage in the Matanuska Valley and on the Kenai Peninsula.

Following are brief descriptions of Alaska's three major grain crops, and three experimental crops, triticale, rye, and speltz.

Barley

Barley, because of its short growing season requirements and ability to grow to maturity in cool temperatures, must be considered the grain most adapted to far-north environments. Only spring barley is important to Alaska; winter varieties lack hardiness and have a low winter survival rate.

The major barley growing region in Alaska is the Tanana Valley, which produced 92 percent of the state's barley crop in 1982. This is due primarily to the Delta Agricultural Project, established by the state in 1978 to help develop Alaska's agricultural resources. Through the project, agricultural rights to about 84,000 acres of land near Delta Junction have been sold for the purpose of barley farming.

Research done on the protein content of Alaska-grown barley has shown that barley grown in arctic and subarctic regions is higher in protein than that grown farther south. This is mainly due to the longer daylight hours, which enable greater amounts of soil nitrogen to be taken up by the plants and converted into grain protein. The protein content can be increased even more by applying nitrogen fertilizers. However, the value of a high-protein feed barley has

Right — *Workers harvest barley from Barney Hollembaek's Delta Junction farm in 1982. The carefully tended farm yielded 75 bushels per acre.* (Cathy Birklid, University of Alaska Agricultural Experiment Station)

Below — *A worker cuts oat hay with a swather on Bob Milby's Matanuska Valley farm.* (Bob Cellers)

A farmer uses a tractor to move rolls of hay down a road in the Matanuska Valley. (Jane Gnass)

Smoke rises as debris is burned off a field being cleared for barley near Delta Junction. (Walt Matell)

A Delta farmer cultivates his field in late summer. By mid-1983, nearly 85,000 acres of land in the Delta Junction-Clearwater area had been sold by the state for the purpose of grain farming. (Alissa Crandall)

not been established, particularly since such grain does not command a higher price at the market. Also, because of the cost of the additional nitrogen, it may be more economical to supplement feed with a less costly and readily available protein source, such as fish by-products.

Livestock feed barley varieties are the choice of most Alaskan farmers, who raise the grain primarily for domestic use. A small amount of barley may be exported in years when crop yields are high. Standard varieties suited to Tanana Valley farming include **Galt, Otra, Weal, Otal,** and **Datal.**

Oats

Oats are considered the second most-adapted grain crop for northern farmers. Although most oat varieties have a longer growing season than barley, they are able to mature in cool temperatures. Another advantage of oats is that they are more tolerant of acid soils than barley or wheat, and can produce high yields in soil with pH ranges of 5.0 to 5.5. Yields during the 1982 season averaged 130 bushels per acre at Delta Junction and 160 bushels per acre at Fairbanks.

Oats are a dual purpose crop in Alaska. They can be harvested before maturity as hay for forage, or harvested at maturity for grain. When harvested for grain, the remaining straw can provide a significant secondary crop. Oats grown for grain must be planted fairly early, preferably in mid- to late May, to allow sufficient time for the crop to mature.

Nip is probably the best all-purpose oat variety for Alaska. It performs well under a wide range of growing conditions, matures very early, produces tall growth, and can be grown for forage. Some farmers prefer Nip because it can be planted almost a week later than other varieties and still reach maturity. Nip seed is available only in Alaska and has been scarce in recent years.

Other standard varieties suited to Alaska include **Pendek, Rodney,** and **Toral.** New varieties are constantly being tested; recently introduced varieties which have proven successful include **Athabasca, Cascade,** and **Pol.**

Wheat

Two species of wheat, commonly called bread wheat and macaroni wheat, are widely cultivated. The bread wheats are subdivided into categories based on their growth habits — spring wheats and winter wheats. Spring wheats show the greatest adaptability to Alaska; winter varieties have poor survival rates and reduced yields. Macaroni wheats have lower yields than bread wheats and require a longer growing season, making them only marginally successful in Alaska.

Wheat has a narrower range of adaptation than barley or oats — it is more sensitive to cool temperatures, and, under ideal weather conditions, matures 10 days later than barley. Thus, when evaluating new wheat varieties, early maturity is of far more importance than yield.

Wheat should always be the first crop planted; seeding should begin as soon as the soil can be tilled in late April or early May to allow it to reach maturity. Standard wheat varieties which have proven successful in the Tanana Valley include **Gasser, Park, Chena, Ingal,** and **Nogal.**

Average wheat yields during the 1982 season were

Large silos mark Alaska Grain Company's storage and fertilizer facility near Delta Junction. (Alissa Crandall)

50 bushels per acre at Delta Junction and 86 bushels per acre at Fairbanks.

Triticale

This grain, described by some as the crop of the future, was grown experimentally in the Tanana Valley in 1980. Triticale is the first man-made cereal grain, the result of a cross between wheat and rye, and is suitable for human or animal consumption. Its protein content averaged 16.8 percent, slightly less than wheat but more than barley and oats. Attempts are being made to develop a triticale with the quality and productivity of wheat and the hardiness of rye.

Crop yields average 3,700 pounds per acre (standard bushel-per-acre figures are not available for triticale), and many varieties produce yields equal to or greater than wheat or barley. Triticale averages 109 days from planting to maturity, compared to 81 days for barley, 99 days for oats, and 97 days for wheat. Because of the long growth requirement, this grain is considered a marginal crop for most areas of Alaska.

Triticale is susceptible to ergot, a fungus disease common to rye which renders the grain toxic to livestock and humans. Agricultural researchers hope that ergot-resistant varieties can be developed.

Rye

Rye can be grown successfully in some areas of Alaska but has never been an important crop. The major limitation to rye production is its susceptibility

A combine stands on the edge of a mixed field of weal barley and oats at the University of Alaska's Agricultural Experiment Station in Palmer. The grains are grown as feed for the station's livestock.
(Shelley Schneider)

Left — *A herd of black Angus and Holstein cattle ready for slaughter are held on the John Rutt farm near Delta Junction. The Delta Agricultural Project has provided an inexpensive source of feed for local livestock.* (*Jill Shepherd, staff*)

Overleaf — *Acres of golden grain grow on an experimental field near Delta Junction. Snowcapped peaks of the Alaska Range rise in the distance.* (*Sharon Paul, staff; reprinted from* The MILEPOST®)

to ergot disease, which makes the grain unusable as food. The highest yielding variety, **Gazelle,** produced an average of 65 bushels per acre in the Agricultural Experiment Station's 1980 testing program.

Speltz

Speltz (also called spelt), a primitive type of wheat that retains its hulls when threshed, is grown as a food crop in extreme environments such as high altitudes in the upper Rhine region of central Europe. It is currently of little economic importance in the United States, although small quantities are grown for use as livestock feed or as a novelty crop.

Tests conducted at Fairbanks and Delta Junction in 1980 indicate that speltz can be grown successfully in the Tanana Valley. The grain was later to mature than wheat, but continued to ripen under cool, wet conditions, and produced respectable yields.

MIN TILLAGE
BARLEY
AFTER RAPESEED
86 55 51
UREA

Livestock

Right — *A three-day-old calf suckles its mother at a farm in Palmer. The cow, a Hereford-Holstein mix, was bred to a Hereford bull to produce this calf.* (Jane Gnass)

Far right — *Herders round up cattle near Swift Creek above Kachemak Bay in fall. These cattle have grazed all summer on the Fox River Flats at the head of the bay.* (Chlaus Lotscher)

Cattle, beef and dairy, have been raised successfully in various parts of Alaska, but the cost of feed and of maintaining animals through harsh winters has hindered industry development. Ranches with beef cattle have been centered on Kodiak Island, on the Kenai Peninsula, in the Aleutians, and in the Delta area of the Interior. On Kodiak, Kodiak Cattle Company runs beef cattle and bison. Unlike cattle, bison do not require food supplements during winter and are therefore cheaper to raise. Delta cattle ranchers depend on nearby barley farmers for much of their cattle feed. But they must contend with severe winters, and they must provide substantial shelter for their stock.

Dairying in Alaska has been given a big boost with the 1983 start of the Point MacKenzie dairy project

Right — *Holsteins graze on Neil Schenk's farm at Delta. Their friend is a registered Holy Cross donkey named Lucinda. (Cathy Birklid, University of Alaska Agricultural Experiment Station)*

Above — *Bob Estelle, assistant herdsman at the University of Alaska Agricultural Experiment Station in Palmer, spends part of his day milking the farm's 150 Holstein dairy cows. After milk production from each cow is tabulated, the warm milk is transferred into these five-hundred-gallon stainless steel cooling and storage tanks before it is shipped to Matanuska Maid dairy for further processing and distribution (Shelley Schneider)*

across Knik Arm from Anchorage. More than 15,000 acres of birch and scrub spruce are being cleared in preparation for growing oats and hay to feed dairy cattle. In addition to Point MacKenzie, dairy farmers in the Matanuska Valley produce milk for the south-central market. One cooperative, Matanuska Maid, sells dairy products in the Anchorage area. For some years, a small dairy operated on the Kenai Peninsula at Kasilof. However, the dairy was forced to shut down its operation because the price of milk declined and the price of hay increased.

A few dairy farms in southeastern Alaska have folded. Competition from shipped-in whole milk with the improved transportation of these times, and also from imported dry milk solids, have made the Juneau, Ketchikan, Wrangell, Petersburg, and Sitka small dairies part of history. Here and there an enterprising bush resident has shipped in an occasional Jersey, but except for along the rail belt and Kodiak, the cow in Alaska is pretty much a thing of the past.

Poultry raising is increasing in Alaska, primarily for eggs. Egg ranchers operate successfully in Seward, in the Matanuska Valley, in the Interior, and on Kodiak. Small ranches at Kenny Lake in the Copper River Valley supply eggs to Valdez, and individuals throughout the state keep small flocks of chickens in their yards.

Commercial sheep raising in the state has centered around the isolated Aleutian Islands in western Alaska. Sheep arrive on the island by boat and are sold to local residents. In previous decades, sheep populations on the islands reached into the thousands. Uncertainty over future ownership of land in the region, and difficulties of reaching markets have led to declining herds. A few farmers throughout the rest of the state raise sheep incidentally to other crops or livestock.

Left — *Karen Lee purchased 70 Holstein dairy cows from farmers in the Matanuska Valley to stock her new dairy in the Point MacKenzie Agricultural Project. The rest of the cows for the 210-staff dairy will be bought from around the state. Like many of the other dairy farmers on the approximately 15,000-acre, state-sponsored project, Lee will supplement the cows' feed with barley grown on her farm.* *(Jill Shepherd, staff)*

Above — *Miss Baldy, a Hereford-Holstein cow, doesn't seem to mind the chickens which roam the grounds of the Gunlogson family's Hillside Boarding Kennel, east of Anchorage in the foothills of the Chugach Mountains.* *(Shelley Schneider)*

Pigs, too, are raised throughout Alaska, but there are few large commercial producers of pork. In Fairbanks, McKee Incorporated raises about one thousand pigs for meat and as weaner pigs for others to raise. The Bannon farm at Delta raises about two thousand pigs for slaughter each year. Perky Pig Farm in Wasilla produces hogs for the southcentral Alaska market.

Farmers in the Matanuska Valley maintain a couple of dairy goat herds. Many smaller herds and individual goats are raised throughout the state. There are no commercial rabbitries in the state, but individuals do raise rabbits.

The ranks of Alaska's livestock also include a few uncommon species such as bison and reindeer. Alaska Department of Fish and Game biologists oversee wild herds of bison in various parts of the state, chiefly at Delta, Farewell, and in the Copper River Valley. Kodiak Cattle Company herds bison on Kodiak Island, and a few bison forage on small islands off the Alaska Peninsula. Individuals, such as Berle Mercer, have tried from time to time to raise bison. Mercer maintains a small herd near Lignite in Alaska's Interior.

Officials estimate there are 12 to 15 reindeer herds in Alaska with a total population ranging between 25,000 and 35,000. These domestic cousins of caribou contribute substantially to the economy in western Alaska, especially on the Seward Peninsula and on the Baldwin Peninsula to the north. Reindeer antlers are clipped each summer for sale to the Oriental market where the antlers are finely sliced or ground into a powder. The end product is sold for medicinal purposes. Reindeer meat is taken for subsistence and sold commercially. Reindeer hides are used for sleeping mats and sled covers. Skin sewers make mukluks from the hide of the leg, and reindeer hooves are sometimes worked into jewelry.

Left — *Scottish Highland and other cattle are herded into an enclosure on Umnak Island in the Aleutians. Substantial cattle and sheep herds have foraged on the island in years past, but management changes and uncertainty over land leases and markets have brought about a decrease in the herds.* (Staff)

Below — *Beef cattle take it easy at their enclosure on Mission Road on Kodiak Island. Most beef is sold directly to the consumer at the end of the grazing season to lessen marketing and transportation problems.* (Perry Valley)

Left — *Scottish Highland cattle, still raised in interior Alaska, are not a common breed. While they are thrifty eaters, they don't have the size or meat of other breeds, and some cattle ranchers maintain their long horns are a drawback. These crossbred Scottish Highland cows, shown here on a small farm near Fairbanks, are owned by Aurora Cattle Company, which also runs the herd of registered Galloway cattle grazing in the pasture in the background.* (Jill Shepherd, staff)

Two young registered Galloway bulls from a herd of about 45 animals peer out from their pen on John Rutt's Delta I Agricultural Project farm near Delta Junction. Aurora Cattle Company, of College, owns the state's only purebred herd of Galloways and boards most of the herd on Rutt's farm. Most Alaska cattlemen use the Galloways as breeding stock for their crossbreeding programs. (Jill Shepherd, staff)

Liz Muth operates a vacuum tool with suction cups which can lift up to 30 eggs at a time at the production facilities for Totem Eggs in Palmer. (Jane Gnass)

Above — *Dust flies as Rhode Island red hens take their daily dirt bath. Although few commercial chicken ranches exist in Alaska, many families raise laying hens, such as these, or meat chickens.* (Janet and Robert Klein)

Left — *Eggs roll out of the washer and on their way to the candler at an egg processor in the Matanuska Valley.* (Jane Gnass)

A gaggle of geese with their goslings cluster at a watering hole in the Matanuska Valley. Chickens make up the majority of the poultry industry in Alaska, but individuals throughout the state raise ducks, geese, and even a few turkeys. (Jane Gnass)

Left — *Toggenburg dairy goats and Suffolk, Columbia, North Country Cheviot, and Columbia/Suffolk sheep graze in a pasture on Rocky and Sue Goodwin's Windy Woods Farm in Palmer. (Shelley Schneider)*

Below — *These Suffolk sheep have just been sheared by Rocky Goodwin at his farm in Palmer. The Goodwins raise chickens and chinchilla rabbits as well as sheep and goats. (Shelley Schneider)*

Rocky and Sue Goodwin check a black fleece for dirt before rolling and sacking it. The Goodwins have the biggest sheep farm in southcentral Alaska. They started raising sheep in 1975 with 2 head and in 1983 had a herd of 47. The Goodwins hope to expand the variety of breeds in their herd, but new varieties must be brought up from the Lower 48 at a cost of about $400 per animal for air freight. (Bob Cellers)

Sue Goodwin inspects a rolled fleece at her farm in Palmer. Alaska does not have a ready market for the wool, and the Goodwins are considering having the wool spun into yarn which they would sell to weavers. Sheep pelts are cured and sold to individuals and to hospitals and rest homes where they aid in preventing bedsores. (Bob Cellers)

Pete Roberts shears a sheep on his Twitter Creek Ranch north of Homer. Roberts sells the thick, lanolin-rich wool to Kenai Peninsula weavers and to a woolen mill which makes the wool into Hudson's Bay-quality blankets which are then sold by Roberts and his wife, Francie. In the fall, Roberts sells lamb meat. (Janet and Robert Klein)

Above — *Gail Mayo's flock of about 25 sheep — mostly Corriedale crossbreeds — are raised primarily for their wool, although some grass-fed lambs are sold off each spring. In addition to her flock of sheep, which is the largest in Fairbanks, Mayo keeps three horses for riding. Her Suffolk ram is a regular ribbon winner at the Tanana Valley Fair.* (Jill Shepherd, staff)

Right — *Terry Miller and his dog herd sheep on the luxuriant tundra of Umnak Island in the Aleutians. Several years ago leases were granted to thousands of acres on the island for raising cattle and sheep. The animals were processed at a facility at Fort Glenn, built at the eastern end of the island as an air base for Dutch Harbor. Recent years have seen a decline in the number of sheep on the island.* (Staff)

Above — *Sandy, a pregnant Yorkshire pig, takes it easy in a kennel east of Anchorage. Sandy is owned by the Gunlogson family, who also raise chickens, rabbits, cows, and horses in addition to boarding their client's animals.* (Shelley Schneider)

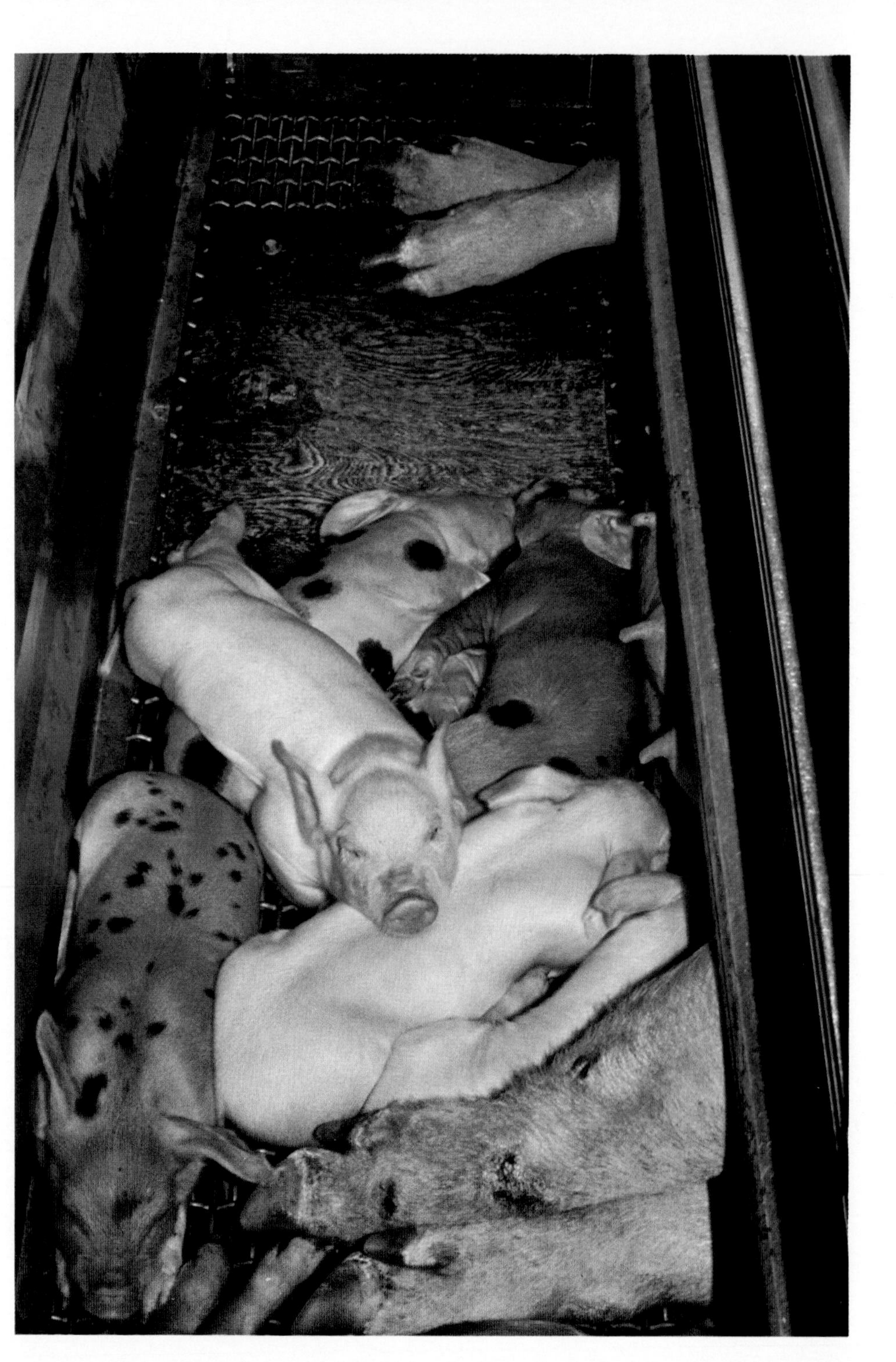

Right — *Newborn to two-week-old piglets sleep in a metal farrowing crate, called a creep. This allows the young piglets to sleep next to but protected from their mother lest she crush them inadvertently. A metal barrier keeps the sow in a separate but open compartment next to her piglets. The average litter size at Perky Pig Farm, where these pigs were raised, is 11 to 12 piglets per sow. After two weeks, the sow and her piglets are moved into a nursery pen where they are kept together for four to five weeks.* (Shelley Schneider)

Weaner pigs work off some excess energy at Perky Pig Farm in Wasilla. (Bob Cellers)

Swine Management Calendar for the Small Farm in Alaska

Breeding Tips

Select a boar for breeding that is at least 8 months old and is healthy, trim, and rugged.

Do not use large, heavy boars for young, first-breeding gilts. This can result in injury to the gilt's back or hind legs due to mounting and riding by a large boar.

Gilts should be about 8 months old and should weigh at least 250 pounds before breeding. Gilts generally reach puberty when they are 6 months old and show signs of estrus (the period of heat, or maximum sexual receptivity). The mature female swine will ovulate (release eggs) every 18-21 days until she is bred and conceives.

The heat period during which a gilt or sow will accept a boar lasts 1-3 days, and ovulation usually occurs from the middle to latter part of the period. Conception rate and litter size may be improved by mating at this time.

Common Swine Breeds

Hampshire	Poland China
Duroc	Berkshire
Yorkshire	Spotted

Crossbreeding two, three, or more breeds of swine is common practice. One objective of crossbreeding is to obtain in the offspring desirable characteristics of the breeds used in the cross. Improved growth rate, litter size, and mothering ability are examples of reasons for crossbreeding in a swine raising program.

GESTATION TABLE
For 114-day gestation period

Breeding Date	Birth Due Date
October 25	February 16
October 30	February 21
November 1	February 23
November 5	February 27
November 10	March 4
November 15	March 9
November 20	March 14
November 25	March 19
November 30	March 24
December 5	March 29
December 10	April 3
December 15	April 8
December 20	April 13
December 25	April 18
December 30	April 23
January 5	April 29
January 10	May 4
January 15	May 9

An easy way to calculate birth due date from breeding date is: 3 months, 3 weeks, 3 days. The average birth date will fall between 112-115 days.

NOVEMBER

Two weeks before breeding swine, increase females' feed intake by 2 pounds per head per day of high energy (low fiber) ration. This is called flushing, and its purpose is to increase the number of ova shed during estrus (heat).

OCTOBER

Obtain a boar and bring him to your farm at least 3 weeks before you intend to use him for breeding to allow him to become familiar with your farm. If you take your female swine to the boar, allow her up to 2 weeks to adjust to her surroundings prior to mating. Be alert to any health problems in the animals.

Do not overfeed a boar so that he becomes fat. This can interfere with his breeding capability.

SEPTEMBER

The optimum weight at which to butcher your hog is 200-240 pounds. Beyond that weight, costs increase as pounds are gained, and most of the added weight is fat. Swine have the highest percentage of dressed carcass weight among farm animals—69-70 percent of live weight. For example, a 200-pound pig should dress out to 140 pounds.

AUGUST

USDA FARMERS BULLETIN Number 2265, "Pork Slaughtering, Cutting, Preserving, and Cooking on the Farm," is available free at your District Cooperative Extension Service Office.

DECEMBER

In Alaska, the author recommends that swine be bred during December, or no earlier than late November. A swine's gestation period (from conception to delivery) is 114 days. (See **Gestation Table** at left.) Plan to have sows or gilts (young female pigs) farrow after mid-March, preferably in April. If they farrow in January or February in periods of cold temperatures and little daylight, more extensive care and equipment is usually required. If you want to raise only one litter of pigs per year to use for food and to sell to friends and neighbors, there is no need to have the sow or gilt farrow earlier than late March or April.

By keeping a boar and sow or gilt separated but close to each other, you will be able to determine when the female's heat period begins by the actions of the boar and sow or gilt. You can then schedule breeding during the latter part of the 3-day heat period when ovulation occurs.

JANUARY

If you own a boar, decide what to do with him after breeding; keep him, sell him, or butcher him for meat. His meat might have an undesirable flavor or odor ranging from slightly distasteful to most objectionable. Castrating a boar several months before slaughtering him will help to reduce the off-flavor risk. However, castration of mature boars drastically shocks the animals, and death may result from excessive bleeding, shock, infection, etc.

Provide the bred sow or gilt with plenty of space for exercise until she is moved to her farrowing quarters. Feed her some distance away from her sleeping quarters.

A heated shelter is usually not necessary. However, do not allow sleeping quarters to be drafty and wet.

FEBRUARY

Scrub down farrowing quarters with disinfectant such as lye mixed with hot water. Use one-half pound lye to 10 gallons water.

MARCH

Move sow or gilt into farrowing quarters 5-7 days prior to expected farrowing date so she will become familiar with her surroundings and recover from nervousness.

Provide her with plenty of fresh water daily.

Scrub sow's or gilt's underline with warm, soapy water to remove manure, dirt, etc. Limit her feed intake 1-2 days after farrowing, and then resume full feed. Keep her pen clean and dry; change its litter and bedding when wet or dirty from manure. Store the manure for your garden.

APRIL

When baby pigs are 1 week old, begin to feed them a 20-22 percent creep (prestarter) feed. Keep them away from drafts. Prevent baby pig anemia.

Castrate baby pigs by 2 weeks of age. Disinfect incision, and make sure baby pigs have a clean, dry pen following the operation.

When pigs are 3-4 weeks old, feed them an 18-20 percent protein starter feed until they are 8-9 weeks old.

Farrowing Tips

Temperature of the hog's farrowing pen should be kept above 50°F. If electricity is available, use heat lamps suspended 24-30 inches above baby pigs during birth.

Immediate temperature surrounding nursing baby pigs should be 80-90°F. Lay thermometer by pigs to determine correct temperature. Baby pigs cannot adjust to temperature fluctuations until they are at least 3 days old.

Be available at farrowing time to give assistance when necessary.

Upon birth, clean mucus from baby pigs' noses and mouths with a clean towel.

Make absolutely sure that each baby pig sucks some of the colostrum milk that a sow initially produces.

Disinfect baby pigs' navels with tincture of iodine soon after birth.

Rail off a corner of the pen with boards, suspend a heat lamp with a chain or wire above the area, and baby pigs will sleep there when not nursing. This will prevent them from being lain on, stepped on, etc.

Baby pigs have 8 (4 pairs) of needlelike teeth. These teeth should be clipped with a side clipper half way to the gum to prevent injury to the sow's teats.

Signs that birth is imminent
Sow will give birth in 1-12 hours when:
she won't eat
she makes a nest of pen litter
milk shows when her teats are squeezed
her vulva is swollen and relaxed
her belly is heavy
she is nervous, shaking, feverish

How to prevent baby pig anemia

Milk is low in iron. Give each baby pig an injection of iron, or swab sow's udder daily with iron sulphate solution.

When the baby pig is 3-4 days old, it should be given an injection of iron in a muscular area such as the ham, neck, or shoulders.

Unless you are experienced in giving shots, choose the ham as the injection site. Use a clean syringe with a needle no longer than one-half inch. The quantity commonly injected is 150-200 milligrams of iron in the form of iron dextran or iron dextrin. One injection usually is enough to sustain the baby pig until it is eating sufficient creep feed to obtain iron.

You can also supply iron to pigs by providing fresh, clean soil for them to root in. Change soil every 2-3 days.

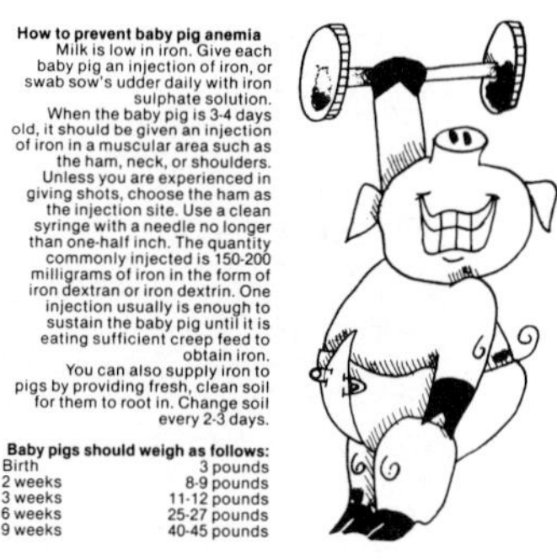

Baby pigs should weigh as follows:

Birth	3 pounds
2 weeks	8-9 pounds
3 weeks	11-12 pounds
6 weeks	25-27 pounds
9 weeks	40-45 pounds

You don't need a scale to weigh a pig

You can determine a pig's weight as follows: Place pig in a normal position as shown in drawing at right. Run tape measure around pig's heart girth (A) to determine its circumference. Measure its back from root of tail (B) to part between ears (C). Then calculate pig's weight with this formula:

$$\frac{\text{heart g.} \times \text{heart g.} \times \text{length}}{400} = \text{weight in pounds}$$

Add 7 pounds to answer if pig weighs less than 150 pounds. No adjustment is necessary if pig weighs 151-400 pounds.

JULY

Always keep swine's sleeping quarters dry with sufficient litter (bedding). Remove manure and store it for use in your garden. Pigs usually will manure a specific portion of a pen or area.

Remove gilts to be used for breeding (replacement gilts) from market (butchering) hogs when they are 4-5 months old or weigh 150-200 pounds. Select gilts that show about average growth rate, length, and muscling. Be sure gilts have at least 12 prominent, well-placed teats and that none are inverted or abnormal. Gilts should have sound feet and legs.

When you have separated gilts from market pigs, feed gilts a nutritionally adequate 14-16 percent protein ration. You do not want them to become fat. Feed them 4-5 pounds of ration daily until 8 months of age. Two weeks before breeding, begin flushing gilts (see **November**).

Consult with your veterinarian or District Extension Agent on recommended drugs for worming pigs. Internal parasites can cost you money by preventing pigs from thriving and reaching their full weight potential. Pigs that fail to thrive are susceptible to disease.

JUNE

Try to keep hogs 500 feet away from human dwellings so that no one will be bothered by the hogs' odors.

Repair, make, or purchase feeding equipment. Concrete troughs are less likely to be tipped over than wooden ones.

Purchase a supply of litter (bedding) for pigs for coming months.

Allocate at least 25 square feet of pen space for each pig that you intend to grow out and fatten.

Provide a well-ventilated shade area (building) for your pigs during hot summer months.

Check your pigs for mange and lice. Consult your District Extension Agent for control measures.

Pigs on good pasture will have sufficient quantities of vitamins A, D, and B-Complex and will require less feed than pigs in confinement on bare ground. Pigs fed in dry lots may need a supplemental source of vitamins. (See **Feeding Tips**.)

MAY

Wean pigs from sow at 6-8 weeks of age. In another 4-4½ months, pigs should weigh between 200-225 pounds and be ready to butcher.

Provide an enclosed shelter (building) to protect pigs from wind, rain, and harsh weather conditions.

A woven wire fence at least 30-36 inches in height is required to confine pigs in a pasture. Keep it in good condition. Remove trash from in and around pigs' living quarters, pasture, or open field area.

Keep stray dogs away from pigs. Dogs will chase them until pigs become exhausted, crippled, or die.

Pigs like to root in the ground with their snout. They can destroy a pasture or dig out from under a fence. To prevent this, you can RING their noses. Ask your District Extension Agent how to do this if you are unfamiliar with the practice.

When pigs have been weaned and the sow or gilt has stopped lactacting and has begun to gain weight, decide what to do with her. You can save her for breeding in late fall, sell her, or butcher her. It is not recommended to breed her again in the summer to have her farrow in the fall. Raising pigs in cold, wet, windy weather requires more effort, time, equipment, feed, etc., than you may want to put into it.

Feeding Tips

The pig's ration from 40-75 pounds should contain 16 percent protein; from 75-125 pounds, 14 percent, and from 125-220 pounds, 12 percent. These rations can be bought commercially or mixed from feedstuffs on the farm.

To mix your own ration, you will need grain, a protein supplement, and a mineral supplement. Barley or corn can be used for grain, soybean meal for protein.

The various fishmeals produced in Alaska from marine products can be used as a portion of the protein supplement. They range in protein from 30-60 percent and can be used for as much as 5 percent of the diet. Barley or corn can be used for the grain, soybean meal for the protein.

The mineral supplement can be provided by mixing together equal parts of steamed bonemeal, ground limestone, and trace mineralized salt. Your District Extension Agent can help you devise a balanced, nutritionally adequate ration for your pig.

Pigs fed under dry lot conditions may need additional vitamins in their rations.

You can add commercially available vitamin premixes or feedstuffs to rations to help supply needed vitamins.

Pigs may be fed by hand twice a day or by using self-feeding equipment to provide a continuous supply of food.

Pigs eat about 2½-3 pounds of feed (dry matter basis) for every 100 pounds of body weight.

How much feed do I need?

A pig will require about 3½-4 pounds of feed per 1 pound of weight gained between 40-220 pounds. A ton of feed is required by 3½ pigs to grow from 40-220 pounds. A pig should gain approximately 1.5-1.8 pounds per day when it is on a growing ration.

Feeding garbage and vegetable crops to swine

Garbage from restaurants, schools, etc., can be fed to pigs as a supplement to a regular feed ration. Do not feed uncooked meat or meat scraps to pigs. It is best to *cook* all garbage before feeding it to pigs. Remove glass and other foreign material from garbage. Garden root crops such as potatoes, turnips, etc., can be used as supplemental feed for hogs, and cooking these crops prior to feeding can increase their nutritive value.

Swine need water every day

35-pound pig	½ gal. a day
200-pound pig	1 gal. a day
Suckling sow	5 gal. a day

Hog Shopping Tips

Swine from University of Alaska Agricultural Experiment Station

As a result of swine research at the University of Alaska's Agricultural Experiment Station in Fairbanks, pigs from the station's swine herd are occasionally available for sale. During the farrowing season in March and April, you can ask the resident animal scientist to have a boar or gilt saved for you to purchase at market value for breeding purposes.

Boredom

Animals can become bored and lonesome. If you intend to purchase a weaned pig, consider acquiring at least 2 pigs. You and a neighbor could jointly raise the pigs, or you could raise a pig for someone else. If pigs are confined, give them an old rubber tire to play with.

Boar is expensive to own

If you have just one or two gilts or sows to breed, it may be an unnecessary cost to buy a boar at 3-4 months of age and feed him until the breeding season. Instead, you can pay a service charge to hire a boar for breeding services. If you do own a boar, you can rent out its breeding services for a fee. You can also butcher a boar for meat. (See **January**.)

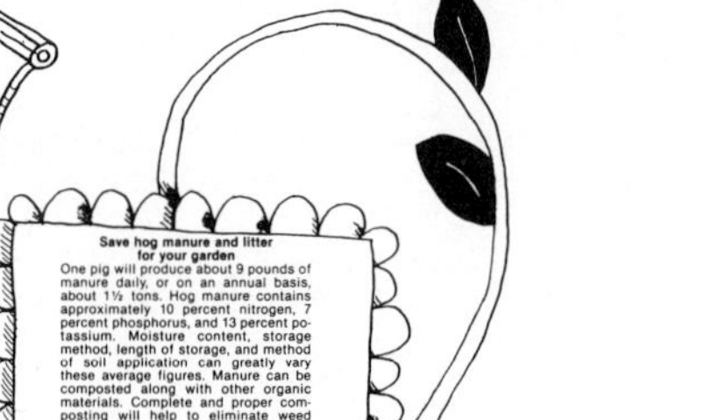

Save hog manure and litter for your garden

One pig will produce about 9 pounds of manure daily, or on an annual basis, about 1½ tons. Hog manure contains approximately 10 percent nitrogen, 7 percent phosphorus, and 13 percent potassium. Moisture content, storage method, length of storage, and method of soil application can greatly vary these average figures. Manure can be composted along with other organic materials. Complete and proper composting will help to eliminate weed seeds that are in raw manure.

(Warren E. Larson, Cooperative Extension Service, University of Alaska)

Below — *Gilt (female pigs prior to their first litter) and boar pigs retained for breeding stock eat out of feed boxes at a farm in the Matanuska Valley. The owner of the farm raises seven pure breeds of swine — Yorkshire, Landrace, Chester White, registered spotted Poland China, registered Duroc, registered Hampshires, and registered Berkshires — in addition to several hybrid varieties.* (Shelley Schneider)

Left — *Terri McKernan removes bacon from a smoker at Perky Packers in Palmer.* (Bob Cellers)

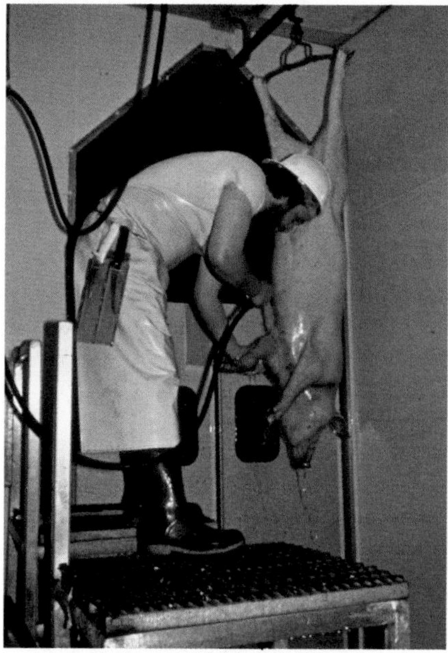

A worker washes a pig before slaughtering at McKee Incorporated, which has the only state-licensed slaughterhouse in the Interior. A family-run business, the farm markets fresh pork, bacon, ham, sausage, and other pork products to stores and restaurants in Fairbanks and delivers to the Parks Highway communities of Nenana, Clear, Healy, and Anderson. The McKees also operate a retail store on the farm they homesteaded in 1957. (Jill Shepherd, staff)

A Toggenburg goat contributes its share to the family larder on this farm in Palmer. Toggenburgs are medium-sized goats of Swiss origin which vary in color from beige to dark chocolate. (Bob Cellers)

Sue Goodwin leads four of her Toggenburg dairy goats out of the barn to a pasture where the goats will graze after their morning milking. (Shelley Schneider)

Deborah Wien raises exotic breeds of rabbits at her rabbitry in North Pole. She trades rabbits with other breeders to expand the number of breeds in her collection. (Cathy Birklid, University of Alaska Agricultural Experiment Station)

These three young rabbits (two New Zealand whites and a Kobuk-New Zealand-Dutch cross) and many other rabbits thrive at the Hillside Boarding Kennel east of Anchorage. (Shelley Schneider)

Left — *Horses graze on lush pasture above Kachemak Bay on the Kenai Peninsula. Officials estimated there were about six thousand horses in Alaska in 1983.* (Chlaus Lotscher)

Right — *Even Alaska has fans of registered Arabians. Gene Grasser displays this stallion, Jamboree Stormy, of which he is part owner, at the Ranch of Envy in Palmer.* (Jane Gnass)

Among the more unusual of Alaska's domestic livestock are reindeer which provide antlers and meat for Alaska's economy. NANA herders and staff are bulldogging these reindeer prior to clipping their antlers at Kotzebue. (Staff; reprinted from ALASKA GEOGRAPHIC®)

A few herds of bison can be found in Alaska, primarily in the Interior and on Kodiak Island. This bison bull is part of the herd kept on Bill Burton's Narrow Cape Ranch in Kodiak. (Walt Them, Cooperative Extension Service)

Three-year-old Shelley Snodgrass gets acquainted with this Holstein-Hereford cow at the Roland Snodgrass farm at Palmer. The herd is owned by John Hett, who leases the land. (Jane Gnass)

Terri McKernan wraps beef roasts at Perky Packers in Palmer. Vosie and Richard Heaton have operated the plant, where they do custom slaughtering, wholesale and retail meat processing, and sausage making, since March 1981. (Bob Cellers)

ALASKA'S LIVESTOCK

	AVERAGE $ VALUE PER ANIMAL:* *Includes all sizes, ages, and sexes of the species.	WHERE:	BREEDS:	COMMENTS/CARE:
CATTLE				
Dairy	$1,500	Matanuska Valley Tanana Valley	Holstein	Calves a few days old can be raised outside in a hutch. Cows usually have their first calves at age two. When ready to produce milk, cows must be brought inside to a controlled environment. Cows can produce milk suitable for human consumption from three to four days after calving to two months before calving. Cows can produce milk until age 18 or 19, but milk production usually tapers off about age 5 or 6. Dairy cattle need fresh water daily, shelter, and feed of hay or silage and high-protein concentrate.
Beef	$200	Delta Kodiak Island and Archipelago Matanuska Valley Kenai Peninsula Aleutian Islands	Galloway Angus Hereford Chianina Scottish Highlander (Most cattle in Alaska are a mixture of several breeds, but there are several small herds of pure breeds.)	Winter feed including hay and silage and concentrates, fresh water daily, and shelter from wind.

	AVERAGE $ VALUE PER ANIMAL:* *Includes all sizes, ages, and sexes of the species.	WHERE:	BREEDS:	COMMENTS/CARE:
POULTRY	$4.70	Seward Kenny Lake Palmer Kodiak North Pole	Leghorn Minorca Plymouth Rock Wyandott Rhode Island red New Hampshire Barred Rock	Must have proper enclosure and even temperature. Ducks, geese, and turkeys are raised in small home flocks throughout the state.
GOATS	$75; $150 for high-quality milking goat	Matanuska Valley Tanana Valley Kenai Peninsula (Individual animals and small groups are raised throughout the state.)	Nubian Toggenburg LaMancha Alpine Saanen (Most goats in Alaska are crossbreeds, but there are small herds of high-quality pure breeds.)	Goats must have winter feed including hay and silage, fresh water daily, and shelter from wind.
RABBITS	No commercial rabbitry in state.	Statewide	Angora New Zealand Californian Chinchilla (Alaska has at least some of all breeds found in the United States.)	Rabbits must be kept dry and out of the wind. Feed commercial rabbit food and some items from the garden. Alaska weather encourages the growth of thick hides which are discounted by commercial users of rabbit hides in the Lower 48.
HOGS	$175	Fairbanks Delta Junction Matanuska Valley	Duroc Yorkshire Landrace	Total confinement in hog house provides best results. Feed barley and a mixture of marine and vegetable protein products.

A white Rhode Island Rock hen and a predominantly black Guinea hen are two of the many chickens raised by the Gunlogson family for eggs and meat at their Chugach foothills home. (Shelley Schneider)

This Black Majestic pig, a hybrid variety which owner Richard Heaton has been developing for four and one-half years, forages near one of the feeding bins at Perky Pig Farm. The farm, with more than fourteen hundred swine, is the only commercial producer of pigs in the Matanuska-Susitna Valley. (Shelley Schneider)

A youngster stops at The Fruit Stand on the Parks Highway to feed a LaMancha and an Alpine dairy goat which the stand owner keeps as pets and as an added attraction for visitors. (Shelley Schneider)

	AVERAGE $ VALUE PER ANIMAL:* *Includes all sizes, ages, and sexes of the species.	WHERE:	BREEDS:	COMMENTS/CARE:
SHEEP	$75	Aleutian Islands Rail belt between Anchorage and Fairbanks	Columbia Corriedale Suffolk	Winter feed, fresh water daily, and shelter from wind. According to some animal scientists, sheep may be the best adapted to Alaska of the country's traditional livestock. Their popularity is increasing among the state's livestock raisers.
HORSES	No figure is available, but horses in Alaska tend to be either high-priced registered stock or grade (mixed breed) pack animals.	Throughout the state except the Arctic.	All breeds	Must have draft-free shelter from the weather and feed of grain and grass hay. Horses will eat barley, but owners usually add supplemental vitamins. Officials placed the state's equine — light horses, ponies, draft horses, mules, and donkeys — population at about 5,600 in 1983.

More information on agriculture in Alaska, from acquiring farmland to learning about new seed varieties, can be obtained from the following agencies:

Agricultural Experiment Station
University of Alaska
Fairbanks, Alaska 99701

Alaska Division of Land
and Water Management
Pouch 7-005
Anchorage, Alaska 99510

Cooperative Extension Service
University of Alaska
Fairbanks, Alaska 99701

U.S. Department of Agriculture
Soil Conservation Service
2221 East Northern Lights Boulevard
Anchorage, Alaska 99508

Note: *Alaska residents should check their telephone directories for the address of their local Cooperative Extension Service or Soil Conservation Service office.*

Aggie Blackmer guides her horse across a shallow braid of the Fox River at the head of Kachemak Bay during a cattle roundup. Horses are used throughout the state for herding. Some even take part in reindeer roundups on the Seward Peninsula. (Janet and Robert Klein)

Alaska Geographic® Back Issues

The North Slope, Vol. 1, No. 1. The charter issue of *ALASKA GEOGRAPHIC®* took a long, hard look at the North Slope and the then-new petroleum development at "the top of the world." *Out of print.*

One Man's Wilderness, Vol. 1, No. 2. The story of a dream shared by many, fulfilled by few: a man goes into the bush, builds a cabin and shares his incredible wilderness experience. Color photos. 116 pages, $9.95

Admiralty . . . Island in Contention, Vol. 1, No. 3. An intimate and multifaceted view of Admiralty: its geological and historical past, its present-day geography, wildlife and sparse human population. Color photos. 78 pages, $5.00

Fisheries of the North Pacific: History, Species, Gear & Processes, Vol. 1, No. 4. The title says it all. This volume is out of print, but the book, from which it was excerpted, is available in a revised, expanded large-format volume. 424 pages. $24.95.

The Alaska-Yukon Wild Flowers Guide, Vol. 2, No. 1. First Northland flower book with both large, color photos and detailed drawings of every species described. Features 160 species, common and scientific names and growing height. Vertical-format book edition now available. 218 pages, $12.95.

Richard Harrington's Yukon, Vol. 2, No. 2. The Canadian province with the colorful past *and* present. *Out of print.*

Prince William Sound, Vol. 2, No. 3. This volume explored the people and resources of the Sound. *Out of print.*

Yakutat: The Turbulent Crescent, Vol. 2, No. 4. History, geography, people — and the impact of the coming of the oil industry. *Out of print.*

Glacier Bay: Old Ice, New Land, Vol. 3, No. 1. The expansive wilderness of Southeastern Alaska's Glacier Bay National Monument (recently proclaimed a national park and preserve) unfolds in crisp text and color photographs. Records the flora and fauna of the area, its natural history, with hike and cruise information, plus a large-scale color map. 132 pages, $11.95

The Land: Eye of the Storm, Vol. 3, No. 2. The future of one of the earth's biggest pieces of real estate! *This volume is out of print,* but the latest on the Alaska lands controversy is detailed completely in Volume 8, Number 4.

Richard Harrington's Antarctic, Vol. 3, No. 3. The Canadian photojournalist guides readers through remote and little understood regions of the Antarctic and Subantarctic. More than 200 color photos and a large fold-out map. 104 pages, $8.95

The Silver Years of the Alaska Canned Salmon Industry: An Album of Historical Photos, Vol. 3, No. 4. The grand and glorious past of the Alaska canned salmon industry. *Out of print.*

Alaska's Volcanoes: Northern Link in the Ring of Fire, Vol. 4, No. 1. Scientific overview supplemented with eyewitness accounts of Alaska's historic volcano eruptions. Includes color and black-and-white photos and a schematic description of the effects of plate movement upon volcanic activity. 88 pages. *Temporarily out of print.*

The Brooks Range: Environmental Watershed, Vol. 4, No. 2. An impressive work on a truly impressive piece of Alaska — The Brooks Range. *Out of print.*

Kodiak: Island of Change, Vol. 4, No. 3. Russians, wildlife, logging and even petroleum . . . an island where change is one of the few constants. *Out of print.*

Wilderness Proposals: Which Way for Alaska's Lands?, Vol. 4, No. 4. This volume gave yet another detailed analysis of the many Alaska lands questions. *Out of print.*

Cook Inlet Country, Vol. 5, No. 1. Our first comprehensive look at the area. A visual tour of the region — its communities, big and small, and its countryside. Begins at the southern tip of the Kenai Peninsula, circles Turnagain Arm and Knik Arm for a close-up view of Anchorage, and visits the Matanuska and Susitna valleys and the wild, west side of the inlet. *Out of print.*

Southeast: Alaska's Panhandle, Vol. 5, No. 2. Explores Southeastern Alaska's maze of fjords and islands, mossy forests and glacier-draped mountains — from Dixon Entrance to Icy Bay, including all of the state's fabled Inside Passage. Along the way are profiles of every town, together with a look at the region's history, economy, people, attractions and future. Includes large fold-out map and seven area maps. 192 pages, $12.95.

Bristol Bay Basin, Vol. 5, No. 3. Explores the land and the people of the region known to many as the commercial salmon-fishing capital of Alaska. Illustrated with contemporary color and historic black-and-white photos. Includes a large fold-out map of the region. *Out of print.*

Alaska Whales and Whaling, Vol. 5, No. 4. The wonders of whales in Alaska — their life cycles, travels and travails — are examined, with an authoritative history of commercial and subsistence whaling in the North. Includes a fold-out poster of 14 major whale species in Alaska in perspective, color photos and illustrations, with historical photos and line drawings. 144 pages, $12.95.

Yukon-Kuskokwim Delta, Vol. 6, No. 1. This volume explored the people and lifestyles of one of the most remote areas of the 49th state. *Out of print.*

The Aurora Borealis, Vol. 6, No. 2. Here one of the world's leading experts — Dr. S.-I. Akasofu of the University of Alaska — explains in an easily understood manner, aided by many diagrams and spectacular color and black-and-white photos, what causes the aurora, how it works, how and why scientists are studying it today and its implications for our future. 96 pages, $7.95.

Alaska's Native People, Vol. 6, No. 3. In this edition the editors examine the varied worlds of the Inupiat Eskimo, Yup'ik Eskimo, Athabascan, Aleut, Tlingit, Haida and Tsimshian. Included are sensitive, informative articles by Native writers, plus a large, four-color map detailing the Native villages and defining the language areas. 304 pages, $24.95.

The Stikine, Vol. 6, No 4. River route to three Canadian gold strikes in the 1800s. This edition explores 400 miles of Stikine wilderness, recounts the river's paddlewheel past and looks into the future. Illustrated with contemporary color photos and historic black-and-white; includes a large fold-out map. 96 pages, $9.95.

Alaska's Great Interior, Vol. 7, No. 1. Alaska's rich Interior country, west from the Alaska-Yukon Territory border and including the huge drainage between the Alaska Range and the Brooks Range, is covered thoroughly. Included are the region's people, communities, history, economy, wilderness areas and wildlife. Illustrated with contemporary color and black-and-white photos. Includes a large fold-out map. 128 pages, $9.95.

A Photographic Geography of Alaska, Vol. 7, No. 2. An overview of the entire state — a visual tour through the six regions of Alaska: Southeast, Southcentral/Gulf Coast, Alaska Peninsula and Aleutians, Bering Sea Coast, Arctic and Interior. Plus a handy appendix of valuable information — "Facts About Alaska." Approximately 160 color and black-and-white photos and 35 maps. 192 pages. Revised in 1983. $15.95.

The Aleutians, Vol. 7, No. 3. Home of the Aleut, a tremendous wildlife spectacle, a major World War II battleground and now the heart of a thriving new commercial fishing industry. Contemporary color and black-and-white photographs, and a large fold-out map. 224 pages, $14.95.

Klondike Lost: A Decade of Photographs by Kinsey & Kinsey, Vol. 7, No. 4. An album of rare photographs and all-new text about the lost Klondike boom town of Grand Forks, second in size only to Dawson during the gold rush. Introduction by noted historian Pierre Berton: 138 pages, area maps and more than 100 historical photos, most never before published. $12.95.

Wrangell-Saint Elias, Vol. 8, No. 1. Mountains, including the continent's second- and fourth-highest peaks, dominate this international wilderness that sweeps from the Wrangell Mountains in Alaska to the southern Saint Elias range in Canada. Illustrated with contemporary color and historical black-and-white photographs. Includes a large fold-out map. 144 pages, $9.95.

Alaska Mammals, Vol. 8, No. 2. From tiny ground squirrels to the powerful polar bear, and from the tundra hare to the magnificent whales inhabiting Alaska's waters, this volume includes 80 species of mammals found in Alaska. Included are beautiful color photographs and personal accounts of wildlife encounters. 184 pages, $12.95.

The Kotzebue Basin, Vol. 8, No. 3. Examines northwestern Alaska's thriving trading area of Kotzebue Sound and the Kobuk and Noatak river basins. Contemporary color and historical black-and-white photographs. 184 pages, $12.95.

Alaska National Interest Lands, Vol. 8, No. 4. Following passage of the bill formalizing Alaska's national interest land selections (d-2 lands), longtime Alaskans Celia Hunter and Ginny Wood review each selection, outlining location, size, access, and briefly describing the region's special attractions. Illustrated with contemporary color photographs. 242 pages, $14.95.

Alaska's Glaciers, Vol. 9, No. 1. Examines in-depth the massive rivers of ice, their composition, exploration, present-day distribution and scientific significance. Illustrated with many contemporary color and historical black-and-white photos, the text includes separate discussions of more than a dozen glacial regions. 144 pages, $9.95.

Sitka and Its Ocean/Island World, Vol. 9, No. 2. From the elegant capital of Russian America to a beautiful but modern port, Sitka, on Baranof Island, has become a commercial and cultural center for Southeastern Alaska. Pat Roppel, longtime Southeast resident and expert on the region's history, examines in detail the past and present of Sitka, Baranof Island, and neighboring Chichagof Island. Illustrated with contemporary color and historical black-and-white photographs. 128 pages, $9.95.

Islands of the Seals: The Pribilofs, Vol. 9, No. 3. Great herds of northern fur seals drew Russians and Aleuts to these remote Bering Sea islands where they founded permanent communities and established a unique international commerce. Illustrated with contemporary color and historical black-and-white photographs. 128 pages, $9.95.

Alaska's Oil/Gas & Minerals Industry, Vol. 9, No. 4. Experts detail the geological processes and resulting mineral and fossil fuel resources that are now in the forefront of Alaska's economy. Illustrated with historical black-and-white and contemporary color photographs. 216 pages, $12.95.

Adventure Roads North: The Story of the Alaska Highway and Other Roads in *The MILEPOST* ®, Vol. 10, No. 1. From Alaska's first highway — the Richardson — to the famous Alaska Highway, first overland route to the 49th state, text and photos provide a history of Alaska's roads and take a mile-by-mile look at the country they cross. 224 pages, $14.95.

ANCHORAGE and the Cook Inlet Basin . . . Alaska's Commercial Heartland, Vol. 10, No. 2. An update of what's going on in "Anchorage country" . . . the Kenai, the Susitna Valley, and Matanuska. Heavily illustrated in color and including three illustrated maps . . . one an uproarious artist's forecast of "Anchorage 2035." 168 pages, $14.95.

Alaska's Salmon Fisheries, Vol. 10, No. 3. The work of *ALASKA®* magazine Outdoors Editor Jim Rearden, this issue takes a comprehensive look at Alaska's most valuable commercial fishery. Through text and photos, readers will learn about the five species of salmon caught in Alaska, different types of fishing gear and how each works, and will take a district-by-district tour of salmon fisheries throughout the state. 128 pages, $12.95.

Koyukuk Country, Vol. 10, No. 4. This issue explores the vast drainage of the Koyukuk River, third largest in Alaska. Text and photos provide information on the land and offer insights into the lifestyle of the people who live and have lived along the Koyukuk. 152 pages, $14.95.

Nome: City of the Golden Beaches, Vol. 11, No. 1. The colorful history of Alaska's most famous gold rush town has never been told like this before. With a text written by Terrence Cole, and illustrated with hundreds of rare black and white photos, the book traces the story of Nome from the crazy days of the 1900 gold rush. 184 pages, $14.95.

NEXT ISSUE

Chilkat River Valley, Vol. 11, No. 3. Its strategic location at the head of the Inside Passage has long made the Chilkat Valley a corridor between the coast and Interior. This issue explores the mountain-rimmed valley, its natural resources, and those hardy residents who make their home along the Chilkat. To members in August 1984. Price to be announced.

The Alaska Geographic Society

Box 4-EEE, Anchorage, AK 99509

Membership in The Alaska Geographic Society is $30, which includes the following year's four quarterlies which explore a wide variety of subjects in the Northland, each issue an adventure in great photos, maps, and excellent research. Members receive their quarterlies as part of the membership fee at considerable savings over the prices which nonmembers must pay for individual book editions.

ALASKA earthlines tidelines

IS ALASKA WARMING UP?

ALASKA EARTHLINES/TIDELINES — an eight-times-a-year newsprint magazine published by The Alaska Geographic Society — deals with a variety of natural resource subjects for classroom study. A new volume begins in September and ends in May. (December/January is a combined issue.) **Student subscriptions** generally include the 8 issues published during a school year. **Single subscriptions** begin with the current issue and continue until 8 consecutive issues have been sent. Subscription prices:

STUDENT: $1.50 per subscription: minimum order, 10 subscriptions sent to one address.

SINGLE: $3.50 per subscription. (Payments to be made in U.S. funds.)

A SAMPLE COPY can be yours for $1.00, postpaid. Make checks payable to The Alaska Geographic Society, and send with your order to *Alaska Earthlines/Tidelines*, Box 4-EEE, Anchorage, Alaska 99509. Your canceled check is your receipt. **GIFT SUBSCRIPTIONS** will be announced to the recipient with a card signed in your name.